UNDERSTANDING SPATIAL MEDIA

UNDERSTANDING SPATIAL MEDIA

EDITED BY
ROB KITCHIN
TRACEY P. LAURIAULT
MATTHEW W. WILSON

Los Angeles | London | New Delhi
Singapore | Washington DC | Melbourne

Los Angeles | London | New Delhi
Singapore | Washington DC | Melbourne

SAGE Publications Ltd
1 Oliver's Yard
55 City Road
London EC1Y 1SP

SAGE Publications Inc.
2455 Teller Road
Thousand Oaks, California 91320

SAGE Publications India Pvt Ltd
B 1/I 1 Mohan Cooperative Industrial Area
Mathura Road
New Delhi 110 044

SAGE Publications Asia-Pacific Pte Ltd
3 Church Street
#10-04 Samsung Hub
Singapore 049483

Editor: Robert Rojek
Editorial assistant: Matthew Oldfield
Production editor: Katherine Haw
Copyeditor: Camille Bramall
Indexer: Rob Kitchin
Marketing manager: Sally Ransom
Cover design: Stephanie Guyaz
Typeset by: C&M Digitals (P) Ltd, Chennai, India
Printed and Bound in Great Britian
by Bell & Bain Ltd, Glasgow.

Library of Congress Control Number: 2016945038

British Library Cataloguing in Publication data

A catalogue record for this book is available from the British Library

MIX
Paper from responsible sources
FSC® C007785
www.fsc.org

ISBN 978-1-4739-4967-6
ISBN 978-1-4739-4968-3 (pbk)

At SAGE we take sustainability seriously. Most of our products are printed in the UK using FSC papers and boards. When we print overseas we ensure sustainable papers are used as measured by the PREPS grading system. We undertake an annual audit to monitor our sustainability.

CONTENTS

LIST OF FIGURES

LIST OF TABLES

CONTRIBUTORS

Michael Batty, Centre for Advanced Spatial Analysis, University College London, London, UK.

Glenn Brauen, Department of Human Geography, University of Toronto, Scarborough, Canada.

Jeremy W. Crampton, Department of Geography, University of Kentucky, Lexington, KY, USA.

Stephen Ervin, Department of Landscape Architecture, Graduate School of Design, Harvard University, Cambridge, MA, USA.

Leighton Evans, School of Arts and Media, University of Brighton, Brighton, UK.

Mary Francoli, School of Journalism and Communication, Carleton University, Ottawa, Canada.

Mark Graham, Oxford Internet Institute, Oxford University, Oxford, UK.

Muki Haklay, Department of Civil, Environmental and Geomatic Engineering, University College London, London, UK.

Rob Kitchin, National Institute for Regional and Spatial Analysis, National University of Ireland, Maynooth, Ireland.

Francisco Klauser, Institut de Géographie, Université de Neuchâtel, Neuchâtel, Switzerland.

Tracey P. Lauriault, School of Journalism and Communication, Carleton University, Ottawa, Canada.

Agnieszka Leszczynski, School of Environment, University of Auckland, Auckland, New Zealand.

Jessa Lingel, Annenberg School for Communication, University of Pennsylvania, Philadelphia, PA, USA.

Gavin McArdle, School of Computer Science, University College Dublin, Dublin, Ireland.

Shannon Mattern, School of Media Studies, The New School, New York, USA.

Harvey J. Miller, Department of Geography, Ohio State University, Columbus, OH, USA.

David Murakami Wood, Department of Sociology, Queens University, Kingston, Canada.

Sung-Yueh Perng, National Institute for Regional and Spatial Analysis, National University of Ireland, Maynooth, Ireland.

Peter L. Pulsifer, National Snow and Ice Data Center, University of Colorado, Boulder, CO, USA.

Britta Ricker, Urban Studies, University of Washington, Tacoma, WA, USA.

Teresa Scassa, Faculty of Law, Carleton University, Ottawa, Canada.

Daniel Sui, Department of Geography, Ohio State University, Columbus, OH, USA.

Jim Thatcher, Urban Studies, University of Washington, Tacoma, WA, USA.

Sarah Widmer, Institut de Géographie, Université de Neuchâtel, Neuchâtel, Switzerland.

Matthew W. Wilson, Department of Geography, University of Kentucky, Lexington, KY, USA.

1

UNDERSTANDING SPATIAL MEDIA

ROB KITCHIN, TRACEY P. LAURIAULT AND MATTHEW W. WILSON

INTRODUCTION

Over the past decade, the practices which produce, process, analyse, share and use digital spatial information have diversified and proliferated. No longer are the handling, storage and examination of digital spatial data confined largely to standalone geographic information systems (GIS), remote sensing packages and specialised geomatics applications that are within the control of a small number of authoritative state, private sector and academic stakeholders, and serviced by a limited pool of skilled personnel. Rather, a varied set of new, networked and often mobile spatial technologies have been developed that are open to use, contributions and editing by anyone with access to the internet. These developments in technology have accompanied rapid shifts in the social, economic, cultural and political geographies of everyday life, with new opportunities for capitalist accumulation and speculation, state and corporate surveillance and governance, and citizen science initiatives.

These new *spatial and locative technologies* include a suite of applications that are explicitly spatial, wherein location and mapping are core to their modus operandi. This includes: online, interactive mapping tools with accompanying application programming interfaces (APIs) that enable the easy production of map mashups which can be embedded on any webpage and push applications beyond desktop GIS (e.g. Google Maps; see Chapters 2 and 3); interactive virtual globes that users can tag and layer data over (e.g. Google Earth); user-generated spatial databases and mapping systems (e.g. OpenStreetMap and WikiMapia, see Chapter 12); augmented spatial media (see Chapter 4); locative media (e.g. satnavs and location-based social networking; see Chapter 5); urban dashboards and citizen-reporting geosystems (see Chapter 7); and geodesign and architectural and planning tools (see Chapter 8). In other cases, applications enable georeferencing that produces spatial (meta)data and can transform the technology into spatial media, but this is not core to its functionality, and the system can operate independently of such spatialisation (see Chapter 6). For example, social media apps such as Twitter and Facebook enable users to georeference tweets/posts, creating a rich set of geosocial data, but the apps work as intended without such georeferencing (Kelley, 2013). Similarly, articles on Wikipedia and in online data repositories can be geotagged, enabling them to be searched by location and spatially visualised. Search has also become spatialised through the location of the searcher. Since 2010, Google has integrated location into all searches either through the internet protocol (IP) address of a computer or the GPS coordinates of a smartphone (Gordon and de Souza e Silva, 2011). Furthermore, maps as branded media are being used to promote institutions and showcase policy and provide a means to navigate web content (see Chapter 18). Concurrently, there are many more non-traditional and administrative datasets making their way into spatial media via open data portals, which in turn are spatialising administrative data (see Chapters 9 and 18).

Geography then has become a key 'organizational logic of the web' and the web has become a key means to mediate space, location and sociality (Gordon and de Souza e Silva, 2011: 3). Indeed, these spatial and locative technologies render virtually everything located or locatable, and thus open to navigation via maps or spatialisations and interpretation through geographical analysis (Gordon and de Souza e Silva, 2011; Wilson and Graham, 2013).

These new technologies have been enabled by the rollout of dense, distributed internetworking – through a variety of communication channels and protocols such as Wi-Fi, bluetooth, Global System for Mobile communication (GSM), Radio-Frequency Identification (RFID), Near-Field Communication (NFC) and the development of enhanced 9-1-1 services. These systems have been extended through ubiquitous computing (computation being accessible through a plethora of networked devices), new mobile platforms with embedded GPS (e.g. smartphones), convergences in media (text, images, maps, audio, video, etc.) and advances in computation, machine learning, indexical and machine-readable identification, non-relational databases and cloud storage (Cartwright et al., 1999, 2007; Taylor, 2005; Crampton, 2009; Kitchin, 2014; Leszczynski, 2014). In particular, the move from Web 1.0 to Web 2.0 in the mid-2000s was instrumental. In the initial rollout of the internet, the web was largely a broadcast medium focused on consumption in which information could be searched, retrieved and read, and services and goods purchased. Spatial information and mapping were largely curated by a few established sites, backed by large capital investment and skilled technical knowledge, that delivered static or dynamic/interactive content through a one-to-many system of communication (such as US and Canada Online National Atlases, Terraserver USA and NASA World Wind, Mapquest; Graham, 2010). These key framework datasets remain a key resource underpinning much spatial media. However, with the shift to Web 2.0, the web became more participatory, social, open (although the extent to which it fulfils these qualities is a continued debate; see Chapter 9), shared and dynamic, with content being produced by users in many-to-many relationships (rather than just specialists), enabled by software infrastructure and APIs that were robust, scalable and, in some senses, invisible to user experience (Kitchin and Dodge, 2011). Web 2.0 facilitated people to communicate and work collaboratively through processes of writing, editing, extending, remixing, posting, sharing, tagging, communicating and so on (Beer and Burrows, 2007). Key developments included the public release of the Google Maps API in 2005 and the centrality of location-awareness in iOS and Android smartphone apps from 2009 onwards that encouraged the development of mobile apps (Crampton, 2009; Gordon and de Souza e Silva, 2011; Kelley, 2013).

Importantly, new networked spatial and locative technologies are not simply a reworking or extension of traditional maps and GIS. Rather, they employ 'different digital structures, techniques and applications', enable different functional and technical affordances, and emerge from different knowledge communities and commercial and political economic contexts (Elwood and Leszczynski, 2013: 549; Wilson and Graham, 2013). As such, while they are related and co-implicated, they are largely 'genealogically distinct from GIS developments' (Leszczynski, 2015: 730; Wilson and Stephens, 2015) and represent 'a profound shift within regimes of the production, dissemination, and institutionalization of geographic information' (Leszczynski, 2012: 72). Moreover, they are much more ubiquitous and entrenched within people's everyday practices than GIS technologies (Leszczynski and Elwood, 2015).

Collectively, these spatial and locative technologies and the effects they engender have been referred to in academia and industry in a number of ways, including the geospatial web or geoweb, neogeography (Turner, 2007), Volunteered Geographic Information (VGI), locative media, spatial media and more specific terms related to certain forms, for example, cybercartography (Taylor, 2005), map hacking (Erle et al., 2005; Schuyler et al., 2005), maps 2.0 (Crampton, 2009), GIS 2.0 (McHaffie, 2008), ubiquitous cartography (Gartner et al., 2007), wikimapping (Sui, 2008), crowd-sourced cartography (Dodge and Kitchin, 2013) and citizen cartography (Graham and Zook, 2013). It is worth untangling and defining each of the more general terms, which are often used interchangeably.

The geospatial web, more commonly known as the *geoweb*, refers to the spatial *technologies* (hardware, software, APIs, databases, networks, platforms, cloud computing), *spatial content* (georeferenced and geotagged data) and the internet-based mapping and location-based *applications/services* that they compose and enable (Scharl and Tochtermann, 2007; Haklay et al., 2008; Crampton, 2009). While the geoweb includes conventional, web-based GIS, it is generally taken to refer to new spatial technologies that are more interactive, participatory, social and generative in nature (Haklay et al., 2008; Kelley, 2013; Elwood and Mitchell, 2013; Wilson, 2014). In essence, the geoweb is the collective noun for the aggregate of spatial technologies and georeferenced information organised and delivered through the internet (Scharl and Tochtermann, 2007; Elwood and Leszczynski, 2011; Leszczynski, 2012). *Locative media* are a subsection of the geoweb that situate users in time and space, and mediate interactions with locations (Wilken and Goggin, 2015). As such, the underlying data, practices and services are location-orientated (Thielmann, 2010). Such locative media include navigation and routing applications, location-based services and advertising practices where users are recommended options with respect to activities based on their present location, and location-based social media (Wilson, 2012). Sui and Goodchild (2011) group the latter into three categories: (1) social check-in sites (e.g. Foursquare); (2) social review sites (e.g. Yelp, Tellmewhere, Groupon);

and (3) social scheduling/events sites (e.g. Meetup). New applications, such as Waze, crowdsource real-time traffic and share navigation recommendations.

Neogeography and VGI refer to the new relations and practices of geographic production and consumption that are created by the rollout and use of the geoweb (Wilson and Graham, 2013). Because the geoweb is largely part of the movement to Web 2.0, 'non-expert' users can use tools to generate, map and share their own spatial data and spatial apps (Turner, 2007; Graham, 2010; Wilson and Graham, 2013; Leszczynski, 2014). In this sense, it constitutes *neogeography* – a new form of producing geography, in that those who interact with and help build the geoweb do so by adding new georeferenced data to initiatives such as OpenStreetMap or WikiMapia, creating map mashups, geotagging encyclopaedia entries, building spatial wikis, reporting urban issues to city geoservices, checking in to locations, etc. Here, geoweb users undertake a form of presumption adding crucial value in the creation of a product or delivery of a service, which they also actively consume, for little or no recompense (Ritzer and Jurgenson, 2010; Dodge and Kitchin, 2013). With respect to a project such as OpenStreetMap, rather than rely on prepared, proprietary and copyrighted cartographic data/products (as created by a national mapping agency such as the UK's Ordnance Survey), users voluntarily collect, clean and upload GPS data, add attribute data, and edit, refine and extend the contributions of others in order to peer-produce a collaborative, detailed, open-source mapping platform (Dodge and Kitchin, 2013; Haklay, 2013). Such spatialised prosumption has been termed *Volunteered Geographic Information* (Goodchild, 2007), though VGI also refers to the generation of spatial information that has not been consciously produced, such as the spatial data fumes of geosocial media (Kelley, 2013; Thatcher, 2014).

Neogeography and VGI, it is thus argued, constitute a new form and era of geographical production/consumption in that control and creation shift from elites and professionals to ordinary people – it is personalised geographical praxis for 'anyone, anywhere, and anytime, and for a variety of purposes' (Haklay, 2013: 56). As such, it is *neo*geography in that the geoweb supersedes and breaks with traditional mapping regimes, practices and technologies, such as conventional cartography and GIS (Leszczynski, 2014). That said, not all of the geoweb is supported by neogeography, with a number of initiatives, especially those supported by the state, relying on more traditional production practices (such as urban dashboards), and the supporting architecture and software being developed by specialist staff. Cybercartography, and more specifically cybercartographic atlases, include participatory mapping, neogeography and VGI, but also reconfigure mapping technology to enable emerging ontologies, especially Indigenous Knowledge representations (see Chapter 13). Further, these atlases recognise that spatial media are also multimodal, and can be multisensory and include multimedia, and that new legal structures are required in order to ensure that collective knowledge represented in maps and atlases, especially Indigenous Knowledge, can be protected in a copyright regime (Taylor, 2005; Taylor and Lauriault, 2014).

Given that the geoweb does not simply present spatial information but mediates a diverse set of socio-spatial practices – communications, interactions, transactions – that extend beyond the representational practices and work of traditional maps, it has been argued that it constitutes a set of *spatial media* (Crampton, 2009; Elwood and Leszczynski, 2013; Wilson, 2014). Early antecedents of this conceptual shift can be found in references to maps/GIS as media (Peterson, 1995; Sui and Goodchild, 2001; Wilson and Stephens, 2015), the spatial mediation of heterogeneous content (Cartwright et al., 2007) and cartographic mediation or the processes of geomediation (Pulsifer and Taylor, 2005). Leszczynski (2015: 729) argues that spatial media refers to 'both the technological objects (hardware, software, programming techniques, etc.) with a spatial orientation' that make up the geoweb, as well as the 'geographic information content forms produced via attendant practices with, through, and around these technologies'. With respect to the latter, spatial media are 'the mediums, or channels, that enable, extend or enhance our ability to interact with and create geographic information online' (Elwood and Leszczynski, 2013: 544). In effect, what the geoweb does is act as spatial *media*; as *interfaces* to create, access and share information and *communication channels* to express spatial relations and meanings (Gordon and de Souza e Silva, 2011; Leszczynski, 2015). From this perspective, the spatial and locative technologies of the geoweb constitute a set of spatial media through which spatial information can be collectively generated, contested, shared and analysed, spatial practices are facilitated, and value leveraged. They are 'sites of potential relations between individuals; persons and places; and people, technology, and space/place'; and they reshape spatial knowledge, mediate spatial behaviour and enact spatial politics (Leszczynski, 2015: 729; Elwood and Mitchell, 2013; Elwood and Leszczynski, 2013). Focusing on the geoweb as media prioritises a concern with the production and flow of information through them, the practices and uses they enable, the work they perform, and the new mediatisations of space, place, location and mobility they enact (Wilson and Stephens, 2015).

In this book, we are concerned with the geoweb, neogeography and spatial media – taken to encompass all of the other neologisms discussed so far – but use spatial media in the title because it both encapsulates the technological components, spatial content (geoweb) and the emergent socio-spatial practices (neogeography), and stresses the work that these do in mediating and conditioning everyday life and producing new spatialities and mobilities. The following section examines some of these new mediatisations and how spatial media are helping to fundamentally transform: the generation of spatial information; the processes and forms of mapping; the nature of space, spatiality and sociality; the practices of mobility and spatial behaviour; the contours of spatial knowledge and imaginaries; and the formation and enactment of knowledge politics.

THE TRANSFORMATIVE EFFECTS OF SPATIAL MEDIA

As documented in detail in Part 3, spatial media have diverse effects on various aspects of everyday life, for example: modifying spatial behaviour (Chapter 16), creating new products and markets (Chapter 17), transforming governance and paradoxically enhancing openness, transparency and participation (Chapter 18), and helping to produce smart cities (Chapter 19), while simultaneously increasing surveillance and control (Chapter 20), and spatial profiling, sorting and prediction

(Chapter 21), as well as transforming the nature of privacy (Chapter 22). Rather than rehearse the arguments presented in these chapters here, it is more instructive to examine how spatial media are transforming thinking with respect to some fundamental geographic and social concepts. Indeed, it is important to stress that spatial media do not challenge and reshape just the practices, discursive regimes and materialities of everyday life, but also how we make sense of them and their affordances and effects.

SPATIAL DATA/INFORMATION

As examined in detail in Part 2, spatial media are inseparable from spatial data, and spatial data/information and the practices that surround them are being transformed alongside general developments in spatial media. First, there has been an explosion in the volume, velocity and coverage of spatial data. Spatial media enable the handling of a diverse set of spatial data, but they also generate massive amounts of such data, including map layers, new framework data (e.g. attribute-rich vector data as in OpenStreetMap), location and movement traces, geotagged and georeferenced data (related to specific phenomena), and metadata (related to posts, comments and photos). Importantly, these data are generated on a continuous basis as spatial and locative media are used, and a much more diverse set of phenomena and practices have gained associated locational data (essentially most activities mediated via the web, especially those using a smartphone or tablet). These data can provide spatial histories of a media and the places and activities captured by them, although it should be noted that because they are generated and stored in proprietary platforms, their long-term preservation is dependent on their host company. Gordon and de Souza e Silva (2011: 19) thus conclude that, given the drive to ensure that all data are geo-referenced as an inherent part of their generation, soon 'unlocated information will cease to be the norm'. In turn, this enables all such data to be tracked and mapped (Thielmann, 2010). This is clearly a significant difference to the pre-spatial media age in which a limited amount of data were spatial, and they were generated on an infrequent basis due to the significant effort and cost expended to produce them. This explosion in production is leading, in the words of Sarah Elwood (2010: 350), to an increasing 'everywhereness' of spatial information in our daily lives.

Second, how spatial data are produced has changed rapidly. Rather than being a skilled process conducted by a limited pool of specialists (e.g. surveyors, GIS technicians, cartographers, spatial database operatives, scientists), usually in the employ of the state or corporations, new actors have become involved. Neogeography, for example, has become a key form of generating spatial data, with data increasingly being generated 'actively/deliberately/knowingly' by millions of ordinary citizens (Graham et al., 2013: 3). This has been accompanied by more automated forms of data production, such as the automatic geotagging of social media posts or the recording of GPS traces as metadata using locative media, in which data are generated 'passively/unconsciously/unknowingly' (Graham et al., 2013: 3). While traditional, formal institutions place a strong focus on standardisation, interoperability and quality/accuracy of spatial data to ensure useable, authoritative and exchangeable data, such an emphasis is variable across spatial media. While some platforms strive to produce spatial data that hold the same qualities as authoritative institutions (e.g. OpenStreetMap vis-à-vis national mapping agencies), in other cases spatial media may be less about scientific and engineered forms of data quality, but more about the qualities of what the data concern and the mapping of narratives (Caquard and Cartwright, 2014). There is also a geography to this production that is highly uneven, largely following the unevenness of physical infrastructure and access to spatial media across the planet, but also censorship regimes and cultural differences in content creation (Graham, 2010). As Graham et al. (2015: 88) note, 'information has always had geography. It is from somewhere; about somewhere; it evolves and is transformed somewhere; it is mediated by networks, infrastructures, and technologies: all of which exist in physical, material places.' Even when spatial data have been produced, there is a geography and politics to their visibility. For example, given that a search is ordered by some criteria (e.g. calculated relevance, popularity) some content is prioritised over others (Graham, 2010). Spatial information then is 'fractured along a number of axes such as location, language, and social networks [and] the resulting constructions of place are complex and far from uniform across space, class, or culture' (Graham and Zook, 2013: 78).

Third, the ontological nature of the data produced is frequently quite different to previous generations of spatial data, often constituting big data or linked data. Big data hold the characteristics of being generated continuously, seek to be exhaustive of a phenomenon or population ($n =$ all), are typically fine-grained and indexical (relating to individual people, places, objects, transactions and interactions) and relational (they can be easily conjoined with other datasets) (Kitchin, 2014; Kitchin and McArdle, 2016; see Chapter 10). Linked data transform the internet from a 'web of documents' to a 'web of data' through the creation of a semantic web that seeks to encode and extract information within web pages – names, addresses, places, product details, facts, figures and so on – through the use of unique identifiers and a markup language to make them visible and enabling others to automatically process, understand and link them together (Berners-Lee, 2009; Miller, 2010; see Chapter 13). While many of the new spatial data being generated are privately held by states or companies, some are open in nature, available to citizens and companies to use (see Chapter 9). The ontological security of spatial big and linked data is unstable due to the continuous and ever-shifting nature of the data generated and the mutability of the underlying technologies and algorithms. As Graham et al. (2013) note, spatial media data are less coherent and fixed, owing to additions, edits, and the contestation and spatial politics of content (e.g. edit wars). Moreover, spatial media themselves have an evolving form, constantly being tweaked and refined, and are designed to provide tailored content based on the profile/location of the user so that there are no fixed representations of place. As such, spatial media and their spatial data 'are enacted and practised in contingent and relational ways', being 'necessarily spatially, temporally and personally context-dependent' (Graham et al., 2013: 467).

MAPPING

Until recently mapping was understood as a representational science; one of producing spatial representations of geographic relationships. Within this conception,

maps sought to faithfully, objectively and accurately capture and portray the absolute position of spatial relations (Robinson et al., 1995). The critique of this notion
was that mapping was far from a neutral exercise and was saturated with power and
ideology (Harley, 1989). In contrast, over the past 15 years or so, mapping has been
reconceptualised within a post-representational perspective, that is, a position that
does not privilege representational modes of thinking (wherein maps are assumed to
be mirrors of the world), nor does it automatically presume the ontological security
of a map as a map (Kitchin, 2010). For example, Del Casino and Hanna (2005) argue
that maps are in a constant state of becoming; that they are 'mobile subjects' whose
meaning emerges through socio-spatial practices of use that mutate with context and
is contested and intertextual. In other words, the map is not fixed at the moment of
creation, but is in constant modification where each encounter with the map produces new meanings and engagements with the world. Similarly, Kitchin and Dodge
(2007: 5) argue that maps are not ontologically secure representations but rather a set
of unfolding practices: '[m]aps are of-the-moment, brought into being through practices (embodied, social, technical), *always* re-made every time they are engaged with;
mapping is a process of constant re-territorialisation. As such, maps are transitory and
fleeting, being contingent, relational and context-dependent' (original emphasis).

While such thinking was initially applied to traditional maps it is clear it has much
resonance for how to make sense of mapping within spatial media. In large part,
this is because spatial media are inherently fluid, transitory, contingent and context-
dependent. While a traditional map gives the impression of a fixity and a totalising
and universal perspective, spatial media are constantly being updated (added to, edited)
and regenerated (e.g. refreshed through zoom, panning, turning on/off features/
layers, during movement), and are contextually filtered in delivery – individually (with
respect to search history), temporally (results change over time), socially (based on social
networks) and geographically (based on present location) (Galloway and Ward, 2005;
Chesher, 2012; Wilson and Graham, 2013; Wilson and Stephens, 2015). As Wilson and
Graham (2013: 6) contend, 'not only do we transduce maps and content in unique,
grounded ways, but the very content that we have available to us varies from person to
person and place to place'. For example, the searching and browsing of a map mashup
of Google Maps and rental and for-sale properties is contextualised with respect to the
user's location and search history and dynamically alters as units are added/removed
from the market. Such contextualisation creates a type of spatial homophily, in which
where we go and what we see is mediated by where and who we are, in turn ensuring
we are spatially and socially sorted to be in places with others like us. With respect to
satnavs, the mapping is aligned to the driver's viewpoint and alters with the real-time
movement of the vehicle in space so that, as the driver navigates, the route and map
are held in alignment (Chesher, 2012). Those that engage with spatial media mappings
are never then simple percipients of maps, but are active in bringing the mappings into
life, shaping their configuration and meanings (Elwood and Leszczynski, 2013; Wilson
and Stephens, 2015).

Indeed, within the context of the geoweb, maps *are* media; they become a prime
communication channel and interface for accessing and revealing web content. As
Gordon and de Souza e Silva (2011: 20) note, 'web mapping is doing more than
transforming mapping practices; it is transforming communication more broadly'.

Mapping is not simply a mode of visualisation, but a 'central organizational device for networked communications', an adaptive interface through which users can access, alter and deploy an expansive database of information, and a platform to socialise spatial information through collective editing, annotations, discussion, etc. (Gordon and de Souza e Silva, 2011: 28). In other words, through its enrolment, the mapping of spatial media content is performing a much more expansive role than revealing spatial relations. In turn, how mappings are being used is becoming a highly immediate, individualised, experiential means to structure search and exploration (not to narrate a set of pre-given spatial meanings), with an approach to asserting credibility based on 'witnessing, peer verification and transparency' (rather than a 'receive and believe' paradigm wherein a map is a secured artefact of legitimacy and authority) (Elwood and Leszczynski, 2013: 554; Wilson and Stephens, 2015). In turn, this is substantially transforming the knowledge politics of mapping (see below).

Further, the relationship between map and territory is being altered. Two of the fundamental conventions of traditional cartography are that space is continuous and ordered and that the map is not the territory but rather a representation of it. As Dodge and Kitchin (2000) illustrated, these conventions are subverted with respect to maps of cyberspace: the spaces of the internet can be discontinuous and organised non-linearly, and in many cases the spaces are their own maps (rather than being external to a representation of data, the map is literally the means to navigate the data). Here, map and territory become synonymous. This is equally becoming the case for spatial media concerning geographic space. Graham et al. (2015: 89) thus contend: 'geographic augmentations are much more than just representations of places: they are part of the place itself; they shape it rather than simply reflect it; and the map again becomes part of the territory.' In other words spatial media do not simply represent space but are integral to the production of space: 'A restaurant omitted from a map can cease to be a restaurant if nobody finds it' (p. 89).

SPACE AND SPATIALITY

A number of commentators have noted that spatial media are transforming the production of space and the nature of spatiality. Spatial media are more and more mediating how space is understood and the interactions occurring within them. Geographic spaces are evermore complemented with various kinds of georeferenced and real-time data – pictures, thoughts, statistics, reviews, historical documents, routes – that can be accessed through a plethora of augmented and location-aware maps and interactive displays that have multiple points of view (Gordon and de Souza e Silva, 2011; Graham and Zook, 2013; de Waal, 2014). This information is observable alongside the space itself at the same time as it generates further data about those places (Chesher, 2012). Moreover, individuals can check into locations, create new georeferenced data, navigate routes, and locate friends and services (de Souza e Silva, 2013). As such, the virtual and material are being entwined, changing the ways in which places are defined and experienced, transforming the 'social production of space and the spatial production of society' (Sutko and de Souza e Silva, 2010: 812; Galloway and Ward, 2005; Graham et al., 2013; de Waal, 2014). For Chesher

(2012), spatial media are shifting the balance in the production of space away from what Lefebvre (1991) termed 'conceived space' (formal abstractions about space such as plans, maps, policy documents) to 'lived space' (space of human action); from representations of space to spaces of representation. In essence, neogeography and access to spatial media open up space for new kinds of engagements and spatial practices, widen a user's sense of perceived space and undermine the centralised power expressed through traditional maps and GIS. In turn, this is leading to the generation of new spatialities and spatial formations that have variously been termed 'code/spaces', 'hybrid spaces', 'digiplace', 'net locality' and 'augmented reality'.

'Code/space' refers to the mutual constitution of software (in this case, spatial media) and the spatiality of everyday life (Dodge and Kitchin, 2005). That is, a dyadic relationship exists between code and spatiality wherein how a space is produced, perceived and experienced is dependent on its mediation through code, and spatial media are dependent on the encoding of spatial relations. Interactions in space mediated by spatial media thus enact a form of code/space. As Kitchin and Dodge (2011) elaborate, the relationship between code and space is neither deterministic (that is, code determines in absolute, non-negotiable means the production of space and the socio-spatial interactions that occur within it) nor universal (that is, such determinations occur in all such spaces and at all times in a simple cause-and-effect manner). Rather how code/space emerges – as with mapping – is contingent, relational and context-dependent. Code/space unfolds in multifarious and imperfect ways, embodied through the performances and (often unpredictable) interactions of individuals and spatial media.

For de Souza e Silva (2006) these code/spaces are hybrid spaces that are simultaneously physical and virtual, a combination of localities and information mediated through spatial media. Such hybridity is evident in the navigation or searching of a locale using mobile locative media, wherein the spatial media directly shape an individual's understanding and experience of a place and, in the case of a location-based social network (LBSN), connections to people in that place (Gordon and de Souza e Silva, 2011). These hybrid spaces, de Souza e Silva (2013: 118) contends, produce 'net locality', namely, 'practiced hybrid space, developed by the constant enfolding of digital information and networked connections into local spaces'. That is, through the use of spatial media an individual is simultaneously local and globally networked. As such, the 'web is brought into the spaces we occupy, and, similarly, those spaces are brought into the web' (Gordon and de Souza e Silva, 2011: 86) and the 'borders between remote and contiguous contexts no longer can be clearly defined' (de Souza e Silva, 2006: 269). For de Waal (2014) this produces both a de-spacing of spatial experience (the ability to share experiences with those not physically present) and an intensification of the same experience through a double interaction (with the space and with absent others). This is leading, he suggests, to a double articulation of place: people meet in a place, such as a shopping mall, discuss the encounter in social media with those present and absent, and keep in contact via social media. In so doing, spatial media heighten the symbolic meaning of spaces.

Zook and Graham (2007: 468) have termed hybrid spaces 'digiplace', noting that the complex entanglements between the physical and virtual are dynamic and mutually constitutive; that is, interdependent. In other words, places are increasingly constituted by a mixture of 'material and virtual social processes and in turn constitute those practices', and individuals navigate such locales using dense clouds of information via spatial media. Given the fluidity, contingency and contextuality of spatial media, locales are revealed as lived, fluid spaces, shaped by space, time, information, user profile, and filtering and framing algorithms (Zook and Graham, 2007). Digiplace is thus a specific and automatically produced spatiality. This spatiality, they have more recently suggested, is a form of augmented reality (Graham et al., 2013; Graham and Zook, 2013; see Chapter 4). They define augmented reality as 'the indeterminate, unstable, context dependent and multiple realities brought into being through the subjective coming-togethers in time and space of material and virtual experience … enacted in specific and individualised space / time configurations' (Graham et al., 2013: 465).

As Leszczynski (2015: 744) notes, such hybridity – whether conceived as net locality or digiplace or augmented reality – means that the experience of spatialities produced by spatial media is always already mediated through the 'the multiple yet momentary comings-together of persons, places, and emergent spatial technologies'. This experience, she argues, is 'intensified by the proximate and synchronous nature of location-aware mobile devices through which this content is both generated and called into being both in situ and in real time' (p. 746). Here, spatiality is recognised as ontogenetic – constantly brought into being – though its articulation is not reducible to technology, social relations or spatiality, but their entanglement (Leszczynski, 2015). Moreover, the new spatialities produced are, in part, a product of new mobilities and spatial practices, but they also facilitate them, inherently reframing the social interactions within spaces and providing different ways to know and navigate locales.

MOBILITY, SPATIAL PRACTICES AND SPATIAL IMAGINARIES

The new spatialities just discussed are the product of new mobilities and spatial practices enabled by spatial media, which in turn are reactive to these spatialities. Spatial media, given their widespread usage and substantive presence in people's daily life (unlike other spatial technologies such as GIS; Leszczynski and Wilson, 2013), increasingly mediate social interactions within spaces and provide different ways to know and navigate locales. For example: satnavs provide calculated routes on dynamically located maps; spatial search and location-based services (LBSs) provide information on and recommendations concerning local businesses; LBSNs enable users to see the real-time location of their friends and to check in to locales; map mashups reveal detailed information about a location; and urban dashboards provide real-time and statistical data visualisations about a place. And, importantly, these tasks can be undertaken in situ, on-the-move and in real time, augmenting a whole series of activities, such as shopping, wayfinding, sightseeing, protesting, etc. In other words, spatial media alter how we understand, relate to, move through, coordinate, communicate in, interact with, and build attachments to space/place. They do this in four ways.

First, as Gordon and de Souza e Silva (2011) note, when using spatial media the perceptual horizon of a person is no longer limited to the environment in which they are located, such as a street, or a limited source of information such as a paper map or guidebook. Instead, the person has access to a range of sources of information, including locative and social media, augmented maps and visualisations, place-related websites and gazetteers, etc. These provide a huge array of supplemental information and filter it with respect to location and activity, which helps guide decision-making and shape spatial practices (Chesher, 2012). As such, Leszczynski (2015: 745) contends that 'everyday encounters with spatial media "actualize new spaces" that are experienced and perceived as interpenetrated – marked, intersected, and constituted' – by spatial data such that 'the experience of *being there* is the experience of being in a location where data is accessible' (Gordon and de Souza e Silva, 2011: 36, original emphasis).

Second, spatial media change the practices of coordination and communication in space, enabling on-the-fly scheduling of meetings and serendipitous encounters (Sutko and de Souza e Silva, 2010; de Souza e Silva, 2013). In the case of LBSN there is no need to actively schedule or make a call, instead viewing the location of friends and intersecting with their location/paths. Sutko and de Souza e Silva (2010: 811) thus suggest that location-aware technologies and the visualisation of spatial relations are replacing the management of time and 'the clock as a medium for coordinating meetings in space'. As such, spatial media demand a rethinking of the processes of sociability (de Waal, 2014). Wilson (2012: 1270) suggests that part of this new sociality is the development of conspicuous mobility created through continuous connectivity to spatial media 'that serves to restructure urban experiences as transactions' by figuring people's mobilities.

Third, at the same time as spatial media can produce serendipitous encounters, they can also work to structure and nudge user perception and movement. For example, suggested routes within a satnav provide a reified path that displaces ad hoc spatial practices (Chesher, 2012). As Chesher (2012: 316) explains, the presented route has 'rhetorical force, with multiple strategies to persuade the driver to take certain paths' and has 'more actuality and force than a street directory flopped open on the passenger seat, and more precision than directions scrawled on a scrap of paper' (p. 323). Likewise the filtering, prioritisation and side-lining of information, for example within an LBS recommender system, works to direct choices (Graham et al., 2013; de Waal, 2014). Indeed, the designers of some spatial media are quite explicit in their desire to generate nudges. For example, Foursquare (an LBSN), states that it is in the 'business of changing user behavior' (Crowley, 2010). Given the commercial nature of most spatial media, it is fair to say that these nudges often have a specific consumption agenda.

Fourth, spatial media help produce new spatial imaginaries. These imaginaries extend well beyond those institutions that have traditionally compiled maps and spatial information. Instead, they are more collective, generative and interconnected, and accessed through a diverse set of apps that provide varying perspectives (Kelley, 2013). They are full of the traces (paths, views, annotations, photos, etc.) of millions of people. These imaginaries can also be highly contested as highlighted by the edit wars in Wikipedia with regards to places (Graham et al., 2015). These imaginaries are 'more than just representations of places: they are part of the place itself; they shape it rather than simply reflect it'; they express attachments to place, but also produce them (Graham et al., 2015: 88). In so doing they also provide a new framework through which identity is formed, constructing an 'inseparable sense of *our-self-our-world*' (Wilson 2014: 536; original emphasis).

KNOWLEDGE POLITICS

A key argument concerning the transformative effects of spatial media is that they radically change the knowledge politics associated with geographic information. Elwood and Leszczynski (2013: 544) detail that 'knowledge politics refers to the use of particular information content, forms of representation or ways of analysing and manipulating information to try to establish the authority or legitimacy of knowledge claims'. Spatial media, it is argued, alter the traditional basis of knowledge politics because they change who is generating spatial data and the nature of expertise and open up different epistemological strategies for asserting 'truth'.

With respect to the former, the advent of neogeography suggests that the production of spatial information has shifted from trained professionals in institutions or corporations to anyone who wants to contribute; from controlled, curated spatial datasets to multivocal, patchwork datasets of curated and volunteered data (Elwood, 2010). As such, there has been a fundamental shift in the processes and power relations of creating and sharing of geographic knowledge, with enhanced access, participation, transparency, and technical literacy and know-how (Elwood, 2010; Haklay, 2013). Some have characterised this move as a form of democratisation, of creating a level playing field, wherein a lay public is able to create, share, explore and interact with maps and other data visualisations (Goodchild, 2007; Turner, 2007; Warf and Sui, 2010; Chesher, 2012). As well as providing an alternative to institutionally curated datasets and tools (e.g. maps, GIS), spatial media can provide challenges to establishment geographies, generating counter-narratives and new knowledge representations as in the case of traditional knowledge (see Chapter 13; Taylor and Lauriault, 2014). In this sense, spatial media are continuing the work initiated within participatory GIS and counter-mapping projects but on a much grander scale (Haklay, 2013). As Elwood and Mitchell (2013) note, neogeography initiatives are thus powerful sites of political action and engagement, and also of political formation, helping to shape the making of political subjects and to mobilise social groups.

Further, the differing technologies and practices of spatial media mean that they are not wholly underpinned by the cartographic and technicist rationalities of GIScience, and they enable different epistemological ways to try and assert legitimacy and authority (Elwood, 2010; Taylor and Lauriault, 2014). In other words, the varying possibilities for structuring, manipulating, sharing and visualising information mean that how knowledge politics is enacted is different (Warf and Sui, 2010; Elwood and Leszczynski, 2013; Wilson and Stephens, 2015). For example, Elwood and Leszczynski (2013: 545) contend that spatial media deploy a variety of geovisual modes to 'structure experiential, exploratory ways of knowing and tend to assert the credibility of those representations through a grounding in practices of witnessing, transparency and peer verification' rather than legitimacy being asserted through 'cartographic abstraction and scientific expertise'. Here, geovisual artefacts 'structure a visual experience' rather than 'narrate a set of pre-given spatial meanings' (Elwood and Leszczynski, 2013: 555). Spatial media also enable other forms of legitimacy, credibility and authoritative knowledge structures to emerge, such as in the case of traditional/Indigenous Knowledge (Pyne and Taylor, 2012), changing normative and legal structures, and providing inclusive mappings (Browne and Ljubicic, 2014; Scassa et al., 2014). Through spatial media the politics of the map/GIS is undermined and replaced with and through a politics of the geovisual/crowdsourcing and new underlying infrastructures which enable these politics to emerge (Wilson and Stephens, 2015; Hayes et al., 2014).

While some spatial media do undoubtedly change spatial knowledge politics there are two challenges to the kinds of changes described above. First, a number of commentators question the extent to which the practices of neogeography are democratising and replacing established, curated geographies (Dodge and Kitchin, 2013; Haklay, 2013). There is an unevenness in the ability to participate due to variance in people's access to the internet, knowledges and skills, with divisions reinscribing traditional divisions along lines of wealth, race, gender and development (Elwood, 2010; Haklay, 2013). Moreover, the affordances of different initiatives are designed, either explicitly or tacitly, to target some groups over others (Leszczynski and Elwood, 2015). Within all initiatives there are hierarchies of participation and control, with commentators such as Carr (2007) asserting that these are necessary to try to assure quality, authority, and usability. No initiative then is either fully democratic or egalitarian, each imbued with circuits of power (Leszczynski, 2014). And, with a few exceptions, such as OpenStreetMap, Wikipedia/Wikimapia and cybercartographic atlases, the underlying technologies, functionalities and governance of spatial media are owned and managed by companies that 'seek to produce new models of capital accumulation by unlocking unwaged virtual labour and information resources and creating new markets' (Dodge and Kitchin, 2013: 20). With respect to Google mashups, for example, Google owns and controls the underlying mapping database, which is professionally sourced, with additional information and mass checking derived from users, and revenue generated via advertising. Google enacts a form of governance that is erratic, opaque, unaccountable and encloses a portion of the geoweb rather than democratising it (Zook and Graham, 2007; Leszczynski, 2012; Scassa, 2013; Saunders et al., 2012). As such, many spatial media do not sit outside of conventional political economic relations (Leszczynski, 2012, 2014; Dodge and Kitchin, 2013).

Second, as discussed in Chapters 15, 17, 20, 21 and 22, it is quite clear that alongside empowering individuals through access to rich information and tools, spatial media also enrol users within new markets and subjugate them within new relations of control and power. While many spatial media are free at the point of use, they have to generate income to cover their costs and produce a profit, and they generally do this through advertising, referrals or selling user data (as many have noted, if the product is free, then the user is the product). Spatial media have radically expanded the volume, range and granularity of the data being generated about people, activities and places, including detailed location and movement tracking, widening the net and scope of surveillance (Elwood and Leszczynski, 2011; Kitchin, 2016). The data generated are easily shared within data markets and can be conjoined with other datasets to extract additional insights, such as predictive profiling, social/spatial sorting and anticipatory governance (Kitchin, 2014). As well as eroding privacy, spatial media and the data they generate are thus being used to shape and regulate behaviour and life chances. As such, a very different set of knowledge politics is being practised to the emancipatory potential envisaged by some.

THE BOOK

Understanding Spatial Media is concerned on the one hand with setting out the nature of spatial media and, on the other, detailing their transformative effects with respect to specific issues. To that end the book has been divided into three interrelated sections. The first section discusses various forms of spatial media, their associated technologies and practices, and issues concerning their operation and how they operate as media. The scope includes GIS, digital mapping, geodesign, social media, locative media, dashboards and augmented reality. The second section concerns the various kinds of spatial data that critically underpin and are generated by spatial media, including geospatial big data, linked geodata, spatial indicators, Volunteered Geographic Information, as well as the data analytics used to make sense of such data. In addition, the section provides an overview of contextual and associated issues, such as open data and legal and policy considerations. The third section focuses on the implications of spatial media and associated spatial data to the practices of living, working, and managing societies and spaces. The scope includes spatial behaviour, business and finance, civic participation, surveillance, spatial profiling, privacy and the creation of smart cities. Each section prioritises different perspectives on spatial media, but they are not mutually exclusive.

Our aim has been to produce a synoptic and critical overview of a phenomenon that has exploded in use and developed rapidly and which has sufficient breadth, depth and reflection to provide a solid understanding of spatial media. The analysis is inherently interdisciplinary, drawing on ideas and work across geography, cartography, sociology, media studies, data science and legal studies. Taken together, we believe the chapters provide a solid foundation for comprehending what spatial media are, how they work, why they matter and how to make sense of them. However, while the text is wide-ranging, it is by no means fully comprehensive. There is a rapidly growing literature seeking to map out and theorise each of the spatial media discussed in the book. And the technologies and their capabilities are ever-evolving, meaning that the book is inevitably a snapshot of a particular time. As a consequence, the book should be read in conjunction with the latest literature in order to follow these rapid changes in developments, thinking and critique.

CONCLUSION

The title of our book is meant to signal our modest goal – to bring together individuals representing key areas of inquiry to discuss spatial media broadly conceived.

While not unproblematic, we hope to both contribute to changing understandings of spatial media and provide a moment of pause, to take stock, to reflect and document this curious moment. No longer entirely comfortable under the subfield of GIScience, digital forms of mapping have become media. While GIScience has also changed rapidly over the past decade it is still largely wedded to a specific set of technologies, practised by a particular set of institutional actors, and rooted in the map as a one-to-many mode of communication model. Instead, spatial media have largely emerged through different technologies and ways of thinking, have a much wider set of corporate, institutional and civic actors, and reframe mapping as interfaces and many-to-many communication channels for accessing, navigating, creating, discussing and sharing information. As such, making sense of spatial media requires an analysis that approaches spatial and locative technologies, the geoweb and neogeography in a much more expansive way than simply adopting a critical GIS perspective. As the chapters in this book make clear, understanding spatial media requires a variety of different perspectives drawn from across the academy – geography, sociology, media studies, computer science, critical data studies, software studies, law, etc. And rather than working in disciplinary isolation, a multidisciplinary approach is required.

As we have argued in this chapter, making sense of spatial media needs to extend well beyond a focus on the spatial and locative technologies themselves and how they work in practice, to consider their implications for how we understand key concepts – spatial data/information, mapping, space/spatiality, mobility/spatial behaviour, spatial imaginaries and knowledge politics. Spatial media impact multiple aspects of social life, including economics, governance, politics and culture, as well as innovation, business, marketing and advertising.

Importantly, then, no longer should spatial media be seen as peripheral to key processes underlying, and key debates about, the formulation and practice of everyday life. Instead, how spatial media have pervaded and are reshaping social, economic and political life needs to be appreciated more widely.

We argue that more work should be focused on situating and unpacking the emergence of spatial media. We agree with Leszczynski and Wilson (2013: 915): 'the rapid proliferation and diversification of spatial media, content forms, and praxes require new empirical, conceptual, and theoretical approaches to apprehend both the nature and implications of these transitions and materialities.' Who stands to benefit from these new innovations? What are the specific uneven topographies of spatial media and associated infrastructures, but also the uneven topographies of access, capital, surveillance and power created in their wake? How are the core underpinning telecomms (e.g. networking) and computing (the cloud, data centres) infrastructure evolving and core framework data being reconfigured? As 'a discursive/material touchpoint for futurity, speculation, and investment', what are the opportunities and limitations for co-optation and resistance to the amassing of capital and the way in which content is or is not volunteered (Wilson, 2012: 1266)? Would we know how to recognise such forms of resistances given our contemporary approaches? How might we situate spatial media 'within historically and geographically contingent enactments of venture capital, the commoditisation of technophilia, networks of natural resource extraction and product disposal, and global divisions of labour' (Wilson and

Graham, 2013: 4–5)? Relatedly, we join numerous social and cultural geographers in the focus on practices, which we suggest requires different approaches. Gillian Rose (2016: 764) has called upon cultural geographers to 'unpack both the symbolism of specific cultural texts but also the production and circulation of those texts by specific forms of media institutions. Similarly, Wilson asks (2014: 536), 'how might we situate the emergence of continuous connectivity as a cultural milieu, and what are the implications for how we study geoweb practices?'

These are just a handful of potential questions that require research and reflection. *Understanding Spatial Media* starts to provide answers to these and related questions. There is clearly, however, much empirical and theoretical work to be done to fill in gaps and provide new conceptual tools and insight. In that sense, this chapter and the book provide an initial grounding with respect to spatial and locative technologies, their effects and emerging debates that will hopefully stimulate and inform further research.

ACKNOWLEDGEMENTS

We would like to thank the contributors to this book for their chapters and their patience, as well as the editorial and production staff at SAGE. Rob and Tracey's research for this chapter and book was funded by a European Research Council Advanced Investigator grant, The Programmable City (ERC-2012-AdG-323636).

REFERENCES

Beer, D. and Burrows, R. (2007) 'Sociology and, of and in Web 2.0: some initial considerations', *Sociological Research Online*, 12 (5). Available at: http://www.socresonline.org.uk/12/5/17.html (accessed 20 July 2016).

Berners-Lee, T. (2009) 'Linked data'. Available at: http://www.w3.org/DesignIssues/LinkedData.html (accessed 20 July 2016).

Browne, T.D.L. and Ljubicic, G. (2014) 'Considerations for informed consent in the context of online, interactive, atlas creation', in D.R.F. Taylor and T.P. Lauriault (eds), *Developments in the Theory and Practice of Cybercartography: Applications and Indigenous Mapping*. Amsterdam: Elsevier. pp. 263–78.

Caquard, S. and Cartwright, W. (2014) 'Narrative cartography: from mapping stories to the narrative of maps and mapping', *The Cartographic Journal*, 51 (2): 101–6.

Carr, N.G. (2007) 'The ignorance of crowds', *Strategy + Business Magazine*, 47: 1–5.

Cartwright, W.E., Peterson, M.P. and Gartner, G. (eds) (1999) *Multimedia Cartography*. Heidelberg: Springer-Verlag.

Cartwright, W.E., Peterson, M.P. and Gartner, G. (eds) (2007) *Multimedia Cartography*, 2nd edition. Heidelberg: Springer-Verlag.

Chesher, C. (2012) 'Navigating sociotechnical spaces: comparing computer games and sat navs as digital spatial media', *Convergence: The International Journal of Research into New Media Technologies*, 18 (3): 315–30.

Crampton, J. (2009) 'Cartography: maps 2.0', *Progress in Human Geography*, 33 (1): 91–100.

Crowley, D. (2010) *Adventures in Mobile Social 2.0: Twelve Months of Foursquare, at Where 2.0*. Santa Clara, CA: O'Reilly Media Inc.

de Souza e Silva, A. (2006) 'From cyber to hybrid: mobile technologies as interfaces of hybrid spaces', *Space and Culture*, 9 (3): 261–78.

de Souza e Silva, A. (2013) 'Location-aware mobile technologies: historical, social and spatial approaches', *Mobile Media and Communication*, 1 (1): 116–21.

de Waal, M. (2014) *The City as Interface: How New Media Are Changing the City*. Rotterdam: nai010 publishers.

Del Casino, V.J. and Hanna, S.P. (2005) 'Beyond the "binaries": a methodological intervention for interrogating maps as representational practices', *ACME: An International E-Journal for Critical Geographies*, 4 (1): 34–56.

Dodge, M. and Kitchin, R. (2000) *Mapping Cyberspace*. London: Routledge.

Dodge, M. and Kitchin, R. (2005) 'Code and the transduction of space', *Annals of the Association of American Geographers*, 95 (1): 162–80.

Dodge, M. and Kitchin, R. (2013) 'Crowdsourced cartography: mapping experience and knowledge', *Environment and Planning A*, 45 (1): 19–36.

Elwood S. (2010) 'Geographic information science: emerging research on the societal implications of the geospatial web', *Progress in Human Geography*, 34: 349–57.

Elwood S. and Leszczynski A. (2011) 'Privacy, reconsidered: new representations, data practices, and the geoweb', *Geoforum*, 42: 5–16.

Elwood, S. and Leszczynski, A. (2013) 'New spatial media, new knowledge politics', *Transactions of the Institute of British Geographers*, 38: 544–59.

Elwood, S. and Mitchell, K. (2013) 'Another politics is possible: neogeographies, visual spatial tactics, and political formation', *Cartographica*, 48 (4): 275–92.

Erle, S., Gibson, R. and Walsh, J. (2005) *Mapping Hacks*. Sebastopol, CA: O'Reilly and Associates.

Galloway, A. and Ward, M. (2005) 'Locative media as socialising and spatialising practices: learning from archaeology'. *Leonardo Electronic Almanac*, MIT Press. Available at: http://www.purselipsquarejaw.org/papers/galloway_ward_draft.pdf (accessed 20 July 2016).

Gartner, G., Bennett, D. and Morita, T. (2007) 'Toward ubiquitous cartography', *Cartography and Geographic Information Science*, 34: 247–57.

Goodchild, M.F. (2007) 'Citizens as sensors: the world of volunteered geography', *GeoJournal*, 69: 211–21.

Gordon, E. and de Souza e Silva, A. (2011) *Net Locality: Why Location Matters in a Networked World*. Malden, MA: Wiley-Blackwell.

Graham, M. (2010) 'Neogeography and the palimpsests of place', *Tijdschrift voor Economische en Sociale Geografie*, 101 (4): 422–36.

Graham, M. and Zook, M. (2013) 'Augmented realities and uneven geographies: exploring the geolinguistic contours of the web', *Environment and Planning A*, 45: 77–99.

Graham, M., Zook, M. and Boulton, A. (2013) 'Augmented reality in the urban environment', *Transactions of the Institute of British Geographers*, 38 (3): 464–79.

Graham, M., De Sabbata, S. and Zook, M. (2015) 'Towards a study of information geographies: (im)mutable augmentations and a mapping of the geographies of information', *Geo*, 2: 88–105.

Haklay, M. (2013) 'Neogeography and the delusion of democratisation', *Environment and Planning A*, 45: 55–69.

Haklay, M., Singleton, A. and Parker, C. (2008) 'Web mapping 2.0: the neogeography of the geoweb', *Geography Compass*, 2 (6): 2011–39.

Harley, J.B. (1989) 'Deconstructing the map', *Cartographica*, 26 (2): 1–20.

Hayes, A., Pulsifer, P.L. and Fiset, J.P. (2014) 'The Nunaliit cybercartographic atlas framework', in D.R.F. Taylor and T.P. Lauriault (eds), *Developments in the Theory and Practice of Cybercartography: Applications and Indigenous Mapping*. Elsevier: Amsterdam. pp. 129–40.

Kelley, M.J. (2013) 'The emergent urban imaginaries of geosocial media', *GeoJournal*, 78: 181–203.

Kitchin, R. (2010) 'Post-representational cartography', *lo Squaderno*, 15: 7–11. Available at: http://www.losquaderno.professionaldreamers.net/wp-content/uploads/2010/02/losquaderno15.pdf (accessed 20 July 2016).

Kitchin, R. (2014) *The Data Revolution: Big Data, Open Data, Data Infrastructures and Their Consequences*. London: Sage.

Kitchin, R. (2016) *Getting Smarter about Smart Cities: Improving Data Privacy and Data Security*, Data Protection Unit, Department of the Taoiseach, Dublin, Ireland. Available at: http://www.taoiseach.gov.ie/eng/Publications/Publications_2016/Smart_Cities_Report_January_2016.pdf (accessed 20 July 2016).

Kitchin, R. and Dodge, M. (2007) 'Rethinking maps', *Progress in Human Geography*, 31 (3): 331–44.

Kitchin, R. and Dodge, M. (2011) *Code/Space: Software and Everyday Life*. Cambridge, MA: MIT Press.

Kitchin, R. and McArdle, G. (2016) 'What makes big data, big data? Exploring the ontological characteristics of 26 datasets', *Big Data and Society*, 3: 1–10.

Kitchin, R., Lauriault, T.P. and McArdle, G. (2015) 'Knowing and governing cities through urban indicators, city benchmarking and real-time dashboards', *Regional Studies, Regional Science*, 2: 1–28.

Lefebvre, H. (1991) *The Production of Space*. Oxford: Blackwell.

Leszczynski, A. (2012) 'Situating the geoweb in political economy', *Progress in Human Geography*, 36 (1): 72–89.

Leszczynski, A. (2014) 'On the neo in neogeography', *Annals of the Association of American Geographers*, 104 (1): 60–79.

Leszczynski, A. (2015) 'Spatial media/tion', *Progress in Human Geography*, 39 (6): 729–51.

Leszczynski, A. and Elwood, S. (2015) 'Feminist geographies of new spatial media', *The Canadian Geographer*, 59 (1): 12–28.

Leszczynski, A. and Wilson, M.W. (2013) 'Theorizing the geoweb', *GeoJournal*, 78: 915–19.

McHaffie, M. (2008) 'GIS 2.0?', formerly available at: http://www.nsgic.org/blog/2008/01/gis-20.html (accessed 30 May 2015).

Miller, P. (2010) 'Linked data and government', *ePSIplatform Topic Report No: 7*. Available at: https://www.europeandataportal.eu/sites/default/files/2010_linked_data_and_government.pdf (accessed 2 August 2016).

Peterson, M. P. (1995) *Interactive and Animated Cartography*. Englewood Cliffs, NJ: Prentice Hall.

Pulsifer, P.L. and Taylor, D.R.F. (2005) 'The cartographer as mediator: cartographic representation from shared geographic information', in, F. Taylor (ed.), *Cybercartography: Theory and Practice*. London: Elsevier. pp. 149–79.

Pyne, S. and Taylor, D.R.F. (2012) 'Mapping indigenous perspectives in the making of the cybercartographic atlas of the Lake Huron Treaty relationship process: a performance approach in a reconciliation context', *Cartographica*, 47 (2): 92–104.

Ritzer, G. and Jurgenson, N. (2010) 'Production, consumption, prosumption: the nature of capitalism in the age of the digital "prosumer"', *Journal of Consumer Culture*, 10 (1): 13–36.

Robinson, A.H., Morrison, J.L., Muehrcke, P.C., Kimmerling, A.J. and Guptil, S.C. (1995) *Elements of Cartography*, 6th edition. New York: Wiley.

Rose, G. (2016) 'Cultural geography going viral', *Social and Cultural Geography*, 17 (6): 763–67.

Saunders, A., Scassa, T. and Lauriault, T.P. (2012) 'Legal issues in maps built on third-party base layers', *Geomatica*, 66 (4): 279–90.

Scassa, T. (2013) 'Acknowledging copyright's illegitimate offspring: user-generated content and Canadian copyright law', in M. Geist (ed.), *The Copyright Pentalogy: How the Supreme Court of Canada Shook the Foundations of Canadian Copyright Law*. Ottawa: University of Ottawa Press. pp. 431–53.

Scassa, T., Taylor, D.R.F. and Lauriault, T.P. (2014) 'Cybercartography and traditional knowledge: responding to legal and ethical challenges', in D.R.F. Taylor and T.P. Lauriault (eds), *Developments in the Theory and Practice of Cybercartography: Applications and Indigenous Mapping*. Elsevier: Amsterdam. pp. 279–97.

Scharl, A. and Tochtermann, K. (2007) *The Geospatial Web: How Geobrowsers, Social Software and the Web 2.0 are Shaping the Network Society*. Dordrecht: Springer.

Schuyler, E., Gibson, R. and Walsh, J. (2005) *Mapping Hacks: Tips and Tools for Electronic Cartography*. Santa Clara, CA: O'Reilly Media.

Sui, D. (2008) 'The "wikification" of GIS and its consequences: or Angelina Jolie's new tattoo and the future of GIS', *Computers, Environment and Urban Systems*, 32: 1–5.

Sui, D. and Goodchild, M. (2001) 'GIS as media?', *International Journal of Geographical Information Science*, 15 (5): 387–90.

Sui, D. and Goodchild, M. (2011) 'The convergence of GIS and social media: challenges for GIScience', *International Journal of Geographical Information Science*, 25 (11): 1737–48.

Sutko, D. and de Souza e Silva, A. (2010) 'Location-aware mobile media and urban sociability', *New Media and Society*, 13 (5): 807–23.

Taylor, D.R.F. (ed.) (2005) *Cybercartography: Theory and Practice*. Amsterdam: Elsevier.

Taylor, D.R.F. (2014) 'Some recent developments in the theory and practice of cybercartography: applications and indigenous mapping', in D.R.F. Taylor and T.P. Lauriault (eds), *Developments in the Theory and Practice of Cybercartography: Applications and Indigenous Mapping*. Elsevier: Amsterdam. pp. 2–16.

Taylor, D.R.F. and Lauriault, T.P. (eds) (2014) *Developments in the Theory and Practice of Cybercartography: Applications and Indigenous Mapping*. Elsevier: Amsterdam.

Thatcher, J. (2014) 'Living on fumes: digital footprints, data fumes, and the limitations of spatial big data', *International Journal of Communication*, 8: 1765–83.

Thielmann, T. (2010) 'Locative media and mediated localities', *Aether: A Journal of Media Geography*, 5: 1–17.

Turner, A. (2007) *An Introduction to Neogeography*. Santa Clara, CA: O'Reilly Media.

Warf, B. and Sui, D. (2010) 'From GIS to neogeogeography: ontological implications and theories of truth', *Annals of GIS*, 16: 197–209.

Wilken, R. and Goggin, G. (2015) 'Locative media – definitions, histories, theories', in R. Wilken and G. Goggin (eds), *Locative Media*. London: Routledge. pp. 1–19.

Wilson, M.W. (2012) 'Location-based services, conspicuous mobility, and the location-aware future', *Geoforum*, 43: 1266–75.

Wilson, M.W. (2014) 'Continuous connectivity, handheld computers, and mobile spatial knowledge', *Environment and Planning D: Society and Space*, 32: 535–55.

Wilson, M.W. and Graham, M. (2013) 'Situating neogeography', *Environment and Planning A*, 45: 3–9.

Wilson, M.W. and Stephens, M. (2015) 'GIS as media?', in S. Mains, J. Cupples and C. Lukinbeal (eds), *Mediated Geographies/Geographies of Media*. Dordrecht: Springer. pp. 209–33.

Zook, M. and Graham, M. (2007) 'Mapping digiPlace: geocoded Internet data and the representation of place', *Environment and Planning B: Planning and Design*, 34: 466–82.

PART 1:
SPATIAL MEDIA TECHNOLOGIES

DE COLORS

SQUARE CHECKINS 2,382.0 / DAY

ER MESSAGES 11,217.0 / DAY

AN HOUSEHOLD INCOME 117,299 $/YEAR

2

GIS

BRITTA RICKER

INTRODUCTION

There are numerous ways in which locational information are both produced and consumed through the use of mobile computers as their users rove the Earth's surface. These phenomena, referred to as spatial media, include but are not limited to location-aware devices, web-based mapping services that facilitate crowdsourcing, and the algorithms used to analyse and distribute location-based information including advertising, navigation, and other (location-based) services and content (Leszczynski, 2015). The locational data from these systems are often visualised through various cartographic representations. These maps hold power and legitimacy, and − with the rise of spatial media − become channels, rather than a mode of communication (Wood, 1992; Elwood and Leszczynski, 2013). Furthermore, recording information about place and then analysing and visualising it can help individuals, organisations, governing bodies and corporations comprehend complex spatial-temporal patterns and make informed decisions about the future (see also Chapter 8).

The tools or applications we use to make these locational decisions extend a form of geographic information systems (GIS). GIS is a broad term that describes a suite of tools, often desktop software but also enterprise systems and processes, including hardware, software, spatial data, procedures, analysis, and outputs (maps) that work together to organise, inventory, display, query and analyse geographic and spatial information (Goodchild, 1997; Longley et al., 2015). The aim of GIS is to help end-users solve a spatial problem, make an informed decision or plan for the future. GIS has played a pivotal role in urban planning, emergency management, environmental engineering, agricultural and forestry management, resource management and, increasingly, the documentation of qualitative information (Weiner and Harris, 2008; Wang et al., 2013). The digital tools within GIS allow users to break down the complexity of the world, by making and documenting observations − abstracting reality − then analysing these abstractions (Brimicombe, 2010). Importantly, I maintain that the core backend spatial databases, algorithms and processes that are required for spatial media are a form of GIS.

Leszczynski (2015) contends that we should consider spatial media as having a distinct epistemology to GIS, blending media and social practices with mapping, and moreover have different genealogies in their development, underpinned by different knowledge domains and industries. Spatial media are seen as new technology, both hardware and software as well as the locational information passing through them, which merge digital content with social practices (Elwood and Leszczynski, 2013; Leszczynski, 2015). I see similarities, more than distinctions, with GIS in this definition of spatial media (although see Figure 2.1). The physical and technological underpinnings of GIS and spatial media are similar: quite simply, they both depend on spatial databases and require geographical knowhow to build the platform required for their operations. As such, spatial media are a form of GIS that permeate our lived experiences and influence our experience in place, which often go unnoticed to a consumer or producer of spatial media. Moments of convergence between GIS and spatial media have created an avenue for a constructive dialogue (Sui and Goodchild, 2011).

The term 'GIS' is often wrongfully equated with proprietary desktop software, such as Environmental Systems Research Institute's (Esri) ArcGIS, which has become an industry standard. Alternatively, there are free and open-source (FOSS) applications such as quantum geographic information systems (QGIS) and R-Spatial. All of these software packages require training and extensive experience to be utilised effectively and were traditionally only accessible to those with requisite financial means or technological expertise. However, new forms of primarily third-party web-based GIS, such as Google Maps, are transforming the use of these tools from expert to everyday, as location is becoming a primary means to engage the web (see Chapter 3; Elwood and Leszczynski, 2013). Increasingly, the type of information that can be included in GIS, as well as those who can participate in their use, are shifting within a spatial media context.

What makes spatial media distinct from other facets of GIS is the focus on crowdsourced data and the inclusion of qualitative information. I suggest that spatial media focus solely on the message generated rather than the expertise and technologies required to build the background processes required to aggregate, analyse and visualise the data (Elwood and Leszczynski, 2013; Thatcher, 2014; Leszczynski, 2015). Nonetheless, the study of spatial media has evolved with close links to critical GIS (Leszczynski, 2015; Wilson and Stephens, 2015), which has sought to critique the technocratic, positivistic nature of GIS (Schuurman, 2000; Elwood, 2006; Thatcher et al., 2015).

The development and study of the use of GIS proceeds as an inherently interdisciplinary field. The processes associated with GIS have traditionally been coupled with physical/natural sciences such as environmental science, geology, surveying and land-use planning. The underlying philosophy of science underpinning these disciplines has meant that GIS has been critiqued for being overly positivistic and only allowing certain epistemological views of the world to be represented (Obermeyer, 1995; Schuurman, 2000). Indeed, those studying the social implications of GIS have

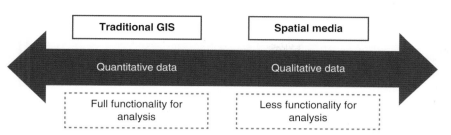

GIS

Traditional GIS | Spatial media

Quantitative data | Qualitative data

Full functionality for analysis | Less functionality for analysis

GIS is more closely tied to quantitative data and information while spatial media are more closely associated with qualitative data and information

FIGURE 2.1 Traditional GIS vs. spatial media

called for GIS/2 (Sieber, 2004), which incorporates qualitative information into GIS, employs more participatory approaches and is framed by other philosophical positions such as feminist thought (Kwan, 1999; Elwood and Cope, 2009).

Now anyone with a global positioning system (GPS) enabled device has the opportunity to contribute geographic information and undertake geographic analysis. This shift in technological availability, information production and geographic computation is profoundly altering the way in which spatial data are being utilised (Leszczynski, 2012; Leszczynski and Wilson, 2013). Trained scientists are no longer the only individuals who can build maps and manipulate spatial data, as these practices have been interwoven into our everyday experiences through the use of spatial media (Elwood and Leszczynski, 2013).

This shift has been made possible by a number of converging factors, most notably the internet, Web 2.0, cloud computing and the plummeting cost of data storage, and advancements in technology and computation. Sui and Goodchild (2001) noted two factors leading to the confluences of GIS and media: first, Esri's efforts to advertise the ability of its GIS products to support communication; and second, digital technologies associated with GIS becoming more pervasive in everyday life. In 2005, Google Earth was released with an application programming interface (API). This gave people with internet access the ability to overlay their own content onto Google Maps with little cartographic or technical expertise. Crampton (2009) documented the rise of the use of these web-based mapping environments and termed these 'new spatial media'.

GIS EVOLUTIONS RELATED TO SPATIAL MEDIA

Table 2.1 provides an overview of noted events associated with the evolution of GIS. Longely et al. (2015) categorise three influential periods: (1) the era of innovation; (2) the era of commercialisation; and (3) the current era of openness and pervasive use. The first two eras tremendously advanced the field of GIS technologically and have been well documented (Brimicombe, 2010; Longely et al., 2015). Digital globes and online interactive mapping tools, such as Google Earth and Google Maps, were the catalyst for this paradigm shift in the third period as mapping tools became

TABLE 2.1 Milestones in the evolution of GIS and spatial media

Year	Significant events associated with the evolution of GIS and spatial media
1996	MapQuest's mapping service is launched, providing driving directions to internet-enabled users
2000	The USA encourages commercial and civilian applications of GPS technology by opening availability of a refined GPS signal
2004	OpenStreetMap is created; all data are voluntarily contributed and open for use and distribution
2005	Google Maps and Google Earth are launched and they release their API. Map mashups emerge
2007	Neogeography book by Turner is published; Goodchild coins VGI. GeoCommons is released as a mapping tool to more easily manipulate census data through a web browser
2008	Google Street View launches. The iPhone, the first touch-screen GPS-enabled smartphone, is released
2009	Quantum GIS is launched
2010	MapBox, MapZen (2013), and many others emerge to deliver software products built around open data, including data hosted by OpenStreetMap
2013	Obama signs an executive order to make open and machine-readable data a requirement of the US government. Civilian use of drones or unmanned aerial vehicles (UAVs) grows significantly; software is developed to process and stitch imagery quickly that are collected from these flying cameras

Adapted from: Longely et al., (2015) and Brimicombe (2010); see these sources for a more comprehensive view of the history of GIS.

more accessible and intuitive to use, enabling more participants (see Chapter 3). At the same time, the USA and European countries, as well as many local government agencies, made efforts to provide more of their (spatial) data to the public through 'open data' initiatives (Sieber and Johnson, 2015; see Chapters 9 and 16). Originally, the UK did not openly share its spatial data with its citizens: one had to pay 'the Crown'. In response, in 2004 Steve Coast started mapping the world with his hand-held global positioning system (GPS) and invited his friends to do the same by using a platform he created at University College London called OpenStreetMap (OSM) (Haklay and Weber, 2008). Users are able to contribute data as well as download them and repurpose them for other projects.

With the ease of use offered by tools like Google Maps and OpenStreetMap, map-user expectations began to shift and, as a result, new mapping tools emerged. For example, Andrew Turner and Sean Gorman created the GeoCommons, which was one of the earliest platforms to give the end-user the ability to spatially analyse US census data in just a few clicks. In 2012, GeoCommons was acquired by Esri and became part of ArcGIS Open Data Portal, a platform designed to work seamlessly with other Esri products for open data to be shared and distributed (http://opendata.arcgis.com/about). In 2009, an open-source alternative to Esri's desktop software was released – QGIS is an official product of the Open Source Geospatial Foundation (OSGeo). Unlike Esri's desktop products, QGIS runs on Unix, Mac OSX and Linux. The majority of plug-ins and tools for QGIS are built by and designed for web developers. QGIS also has many tools that make it easy to render data in traditional or complex GIS formats and export them into more web-friendly formats for more people to view spatial data and information on the web.

These initiatives reveal two significant differences between some spatial media and GIS. Within traditional GIS, software and data are often proprietary, meaning they are owned by a company or institution which retains rights with respect to their use (although Esri users can now share their data via the ArcGIS Open Data platform). Alternatively, open-source software can be free of cost ('free as in beer') and/or be free to manipulate the code ('free as in speech'), and the data these tools render may or may not be open spatial data (see Chapter 9). Irrespective of whether the data are open or proprietary, the energy and innovation necessary to produce spatial data make them a valuable commodity. Initially with the advent of traditional GIS only a handful of experts collected spatial data and produced map outputs; this was later followed by crowdsourced spatial-data-collection projects that relied on the masses to undertake the same tasks (Dodge and Kitchin, 2013; see Chapter 12). However, with respect to wider processing and analysis, the tools afforded to the end-user are typically very limited within spatial media.

EMERGENT ACTORS, FIELDS AND FUTURES

The most influential industrial player in the realm of GIS has historically been and continues to be Esri. Since its inception, Esri has worked closely with industry, governments and academia. Esri has a legacy of hiring experts in the field of cartography, system science, spatial statistics, biology and other fields that stand to benefit from GIS, to work hand in hand with the developers and database administrators to best design products for a range of spatial inquiry. Based on rapid changes in spatial media, Esri is increasingly focusing its development efforts on building web-based tools for its users. However, Esri is confronted with new actors in this rapidly changing field, such as MapBox, MapZen (research division of Samsung), Stamen, Carto and MapSense to name a few, which have built their work on top of the OpenStreetMap base map as a way to circumvent the need to digitise and update their own spatial datasets.

While the use of the term 'spatial media' is uncommon among GIScientists, these researchers seek to address the challenges associated with processing large datasets, such as those generated by social media that have locational information attached (see Chapters 4, 6, 10 and 12). Traditional desktop GIS may not have the capabilities to analyse increasingly large datasets, and as a result, cyberinfrastructures are being created and utilised to help with these tasks. Cyberinfrastructure consists of distributed yet connected computer systems, data storage systems, instruments and data repositories connected through software and high-performance networks to improve research productivity that would not be possible without it. With respect to spatial data, the integration of cyberinfrastructure, GIS, and spatial analysis and modelling is termed 'CyberGIS' (Wang and Armstrong, 2009; Wang et al., 2013).

Governments also have a mandate to collect spatial data for a variety of reasons including public infrastructure management and maintenance, environmental monitoring, fire and police reporting and to meet legislative mandates, to name a few. Traditionally, governments have been primarily concerned with managing these data internally, but the open data movement has placed pressure on governments to present these data to the public (see Chapter 9). The hope is that by opening data, this may lead to increased public participation and engagement with government through data collection and sharing of data through the open data movement (Roche, 2014; Sieber and Johnson, 2015; see Chapters 9 and 16).

With these relatively new capabilities afforded by GIS, more opportunities to participate in GIS are emerging. When we use our smartphones, we create a trace of digital bread crumbs, or 'fumes' (Thatcher, 2014), which are valuable to businesses and governments, as well as to our friends and social networks (see Chapter 5). Governments often use this type of information, for among other things, 'predictive policing' or planning where to position police and medical services during special crowd events. Business analysts are also quick to collect these data and use them to target products to potential buyers via the resale of app-based data. These services are shifting our expectations of business practices. In addition, among friends, people are negotiating social acceptance of spatial media in their interaction with one another. On the one hand, people are proud to show off where they have been, through the use of photo-sharing apps like Instagram and Snapchat, as a kind of conspicuous mobility (Wilson, 2012). On the other hand, some find it 'creepy' to watch each others' and their own movement on a map (Ricker et al., 2015). Furthermore, users may feel discomfort related to who is in control of the data that are produced, and how these data are used for unintended purposes (Elwood and Leszczynski, 2011). For these reasons new research has examined locational masking techniques (Clarke, 2016).

While the social and cultural practices around spatial media are shifting and intensifying, the geospatial technologies, associated tools and data structures are also constantly evolving. As a result, it can be difficult to keep up with the map package and web JavaScript libraries du jour (Roth et al., 2014). From a more traditional GIScience perspective, there are primarily two streams of users who have different expectations – the professional user and the general public – and these end-users may expect different mapping products. However, from a spatial media perspective, we increasingly must also consider the programmers who generate the algorithms running behind the scenes, who set up the databases where the data are stored, and who connect tools and data through different APIs. This is causing challenges for those who teach GIS: how do we teach this rapidly expanding and diversifying field?

As a result of these changes in GIS and spatial media, there are seemingly endless advanced topics associated with the study and use of these technologies. As they evolve their social acceptance, norms and expectations shift and this will provide more opportunities for the use of GIS to grow. Figure 2.2 presents four phases of GIS data flow. In each phase, one can consider the rapid changes brought about by the expanded role of spatial media in everyday life – which involves, generally speaking, more

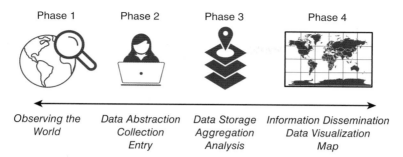

Phase 1   Phase 2   Phase 3   Phase 4

Observing the   Data Abstraction   Data Storage   Information Dissemination
World   Collection   Aggregation   Data Visualization
                    Entry   Analysis   Map

FIGURE 2.2   Phases of data flow associated with GIS and spatial media

efficient or more accurate or more diverse data handling and processing strategies. For instance, in phase 1, data formats are typically either raster (images that are made up of pixels or cells) or vector data (points, lines and polygons). Traditionally, raster data are (tiled) images or other forms of remotely sensed data that form a base map. These datasets are typically expensive to access, requiring images from a plane or a satellite. As demands for better quality and higher temporal and/or spatial resolution increase, the imagery becomes more expensive. However, as the price of unmanned aerial systems (UASs) continues to drop, and their ease of use continues to improve, there will be increasing opportunity for raster data to act as spatial media.

In phase 2, the focus has traditionally been on data collection, entry, and techniques for abstraction and generalisation. New challenges in this phase relate to new digital methods for how we collect and input spatial data into GIS and associated spatial media systems, as well as the use of 'big data' (see Chapters 10 and 14). Big data are extremely large, heterogeneous and often-messy datasets that are associated with multiple forms of data collection, including from social media and the tracking of consumer behaviours. Even those mobile apps that do not seem related to spatial media (e.g. puzzles and other popular games and search engines) are often collecting spatial information about their users (Jung et al., 2012). This leads to significant ethical questions (see Chapters 20, 21 and 22).

As GIS and spatial media begin to intersect, phase 3 activities like data storage, aggregation and analysis will be revolutionised to support activities like the mining of big data. However, the decisions we make to analyse these datasets deeply influence the significance of such analyses (Monmonier, 1996; MacEachren, 2004; Crampton, 2010). In other words, data do not speak for themselves – the way we collect and analyse data influences their meaning (Harley, 1989; Graham and Shelton, 2013).

The final phase is associated with information dissemination and visualisation. This is perhaps where GIS most clearly blends into forms of media. The meanings that maps espouse are strongly related to usability and user experience. Here, it is important to recognise that the digital divide remains an issue and affects access and participation, particularly around the contribution of VGI (Haklay, 2013; see Chapters 4 and 12). Furthermore, research in usability examines the ease with which users can achieve the intended goal associated with a software system (Haklay, 2010).

In addition to these changes in the practices of GIS and spatial media, many scholars continue to draw attention to the social and political implications of these technologies. Elwood (2014) summarises these key transformations resulting from the permeation of the geoweb into our daily lives, including privacy and surveillance (see Chapters 20, 21, 22), and activism and civic engagement (see Chapter 12). GIS and spatial media researchers must maintain a keen awareness of these issues so as not to reproduce inequalities. Those who have access to technology can contribute, but the context in which they are situated (work versus play) also influences the content being collected (Thatcher, 2013, 2014). The views of women and other minorities are not well documented and are thus under-represented in GIS (Stephens, 2013) and spatial media (see Chapter 4). Further, GIS can be used to illuminate inequality and disparities between communities (Kwan, 1999; Schuurman, 2006; Elwood, 2008).

In conclusion, while recent literature suggests that GIS provides an outmoded framework for understanding recent changes in geographic technologies and practices, we must recognise that spatial media are built in part upon the technological scaffolding put forth by traditional GIS. Doing so enables us to draw upon the energies and efforts (and progress) made by interventions created by critical GIS. By acknowledging the significant social, ethical and political interruptions created by GIS and spatial media, we might create a richer, more inclusive and safer environment for those involved (knowingly or unknowingly) with the creation and analysis of our spatial-mediated future. Therefore, it is vital that scholars constantly apply and critique existing usage and development of GIS and its forms within spatial media.

REFERENCES

Brimicombe, A. (2010) *GIS, Environmental Modeling and Engineering*, 2nd edition. Boca Raton, FL: CRC Press/Taylor & Francis Group.

Clarke, K. (2016) 'A multiscale masking method for point geographic data', *International Journal of Geographical Information Science*, 30 (2): 300–15.

Crampton, J. (2009) 'Cartography: maps 2.0', *Progress in Human Geography*, 33 (1): 91–100.

Crampton, J. (2010) *Mapping: A Critical Introduction to Cartography and GIS*. Oxford: Wiley-Blackwell.

Dodge, M. and Kitchin, R. (2013) 'Crowdsourced cartography: mapping experience and knowledge', *Environment and Planning A*, 45 (1): 19–36.

Elwood, S. (2006) 'Critical issues in participatory GIS: deconstructions, reconstructions, and new research directions', *Transactions in GIS*, 10 (5): 693–708.

Elwood, S. (2008) 'Volunteered Geographic Information: future research directions motivated by critical and participatory and feminist GIS', *GeoJournal*, 72 (2): 173–83.

Elwood, S. (2014) 'New spatial technologies, new social practices: a critical theory of the geoweb', *Erlanger Vortrag Zur Kulturgeographie*, 60: 1–6.

Elwood, S. and Cope, M. (2009) 'Qualitative GIS: forging mixed methods through representations, analytical innovations, and conceptual engagements', in M. Cope and S. Elwood (eds), *Qualitative GIS: A Mixed Methods Approach*. London: Sage. pp. 1–12.

Elwood, S. and Leszczynski, A. (2011) 'Privacy, reconsidered: new representations, data practices, and the geoweb', *Geoforum*, 42 (1): 6–15.

Elwood, S. and Leszczynski, A. (2013) 'New spatial media, new knowledge politics', *Transactions of the Institute of British Geographers*, 38 (4): 544–59.

Goodchild, M. (1997) 'What is Geographic Information Science? NCGIA Core Curriculum in GIScience'. Available at: http://www.ncgia.ucsb.edu/giscc/units/u002/u002.html (accessed 20 July 2016).

Graham, M. and Shelton, T. (2013) 'Geography and the future of big data, big data and the future of geography', *Dialogues in Human Geography*, 3 (3): 255–61.

Haklay, M. (ed.) (2010) *Interacting with Geospatial Technologies*. Chichester: John Wiley.

Haklay, M. (2013) 'Neogeography and the delusion of democratisation', *Environment and Planning A*, 45 (1): 55–69.

Haklay, M. and Weber, P. (2008) 'OpenStreetMap: user-generated street maps', *IEEE Pervasive Computing*, 7 (4): 12–18.

Harley, J.B. (1989) 'Deconstructing the map', *Cartographica*, 26 (2): 1–20.

Jung, J., Han, S. and Wetherall, D. (2012) 'Enhancing mobile application permissions with runtime feedback and constraints', in *Proceedings of the Second ACM Workshop on Security and Privacy in Smartphones and Mobile Devices*. New York: ACM. pp. 45–50.

Kwan, M.P. (1999) 'Gender and individual access to urban opportunities: a study using space–time measures', *The Professional Geographer*, 51 (2): 210–27.

Leszczynski, A. (2012) 'Situating the geoweb in political economy', *Progress in Human Geography*, 36 (1): 72–89.

Leszczynski, A. (2015) 'Spatial media/tion', *Progress in Human Geography*, 39 (6): 729–51.

Leszczynski, A. and Wilson, M.W. (2013) 'Theorizing the geoweb', *GeoJournal*, 78 (6): 915–19.

Longley, P., Goodchild, M., Maguire, D. and Rhind, D. (2015) *Geographic Information Science and Systems*, 4th edition. Chichester: Wiley.

MacEachren, A.M. (2004) *How Maps Work: Representation, Visualization and Design*. New York: Guilford Press.

Monmonier, M. (1996) *How to Lie with Maps*. Chicago, IL: University of Chicago Press.

Obermeyer, N. (1995) 'The hidden GIS technocracy', *Cartography and Geographic Information Systems*, 22 (1): 78–83.

Ricker, B., Schuurman, N. and Kessler, F. (2015) 'Implications of smartphone usage on privacy and spatial cognition: academic literature and public perceptions', *GeoJournal* 80 (5): 637–52.

Roche, S. (2014) 'Geographic Information Science I: why does a smart city need to be spatially enabled?', *Progress in Human Geography*, 38 (5): 703–11.

Roth, R.E., Donohue, R.G., Wallace, T.R., Sack, C.M. and Buckingham, T.M.A. (2014) 'A process for keeping pace with evolving web mapping technologies', *Cartographic Perspectives*, 78 (78): 25–52.

Schuurman, N. (2000) 'Trouble in the heartland: GIS and its Critics in the 1990s', *Progress in Human Geography*, 24 (4): 569–90.

Schuurman, N. (2006) 'Formalization matters: critical GIS and ontology research', *Annals of the Association of American Geographers*, 96 (4): 726–39.

Sieber, R.E. (2004) 'Rewiring for a GIS/2', *Cartographica*, 39 (1): 25–39.

Sieber, R.E. and Johnson, P. (2015) 'Civic open data at a crossroads: dominant models and current challenges', *Government Information Quarterly*, 32 (3): 308–15.

Stephens, M. (2013) 'Gender and the geoweb: divisions in the production of user-generated cartographic information', *GeoJournal*, 78: 981–96.

Sui, D. and Goodchild, M.F. (2001) 'GIS as media?', *International Journal of Geographical Information Science*, 15 (5): 387–90.

Sui, D. and Goodchild, M.F. (2011) 'The convergence of GIS and social media: challenges for GIScience', *International Journal of Geographical Information Science*, 25 (11): 1737–48.

Thatcher, J. (2013) 'From volunteered geographic information to volunteered geographic services', in D. Sui, S. Elwood and M. Goodchild (eds), *Crowdsourcing Geographic Knowledge: Volunteered Geographic Information (VGI) in Theory and Practice*. Dordrecht: Springer. pp. 161–74.

Thatcher, J. (2014) 'Living on fumes: digital footprints, data fumes, and the limitations of spatial big data', *International Journal of Communication*, 8: 1765–83.

Thatcher, J., Bergmann, L., Ricker, B., Rose-Redwood, R., O'Sullivan, D., Barnes, T. J. and Young, J.C. (2015) 'Revisiting critical GIS', *Environment and Planning A*, 48 (5): 815–24.

Wang, S. and Armstrong, M.P. (2009) 'A theoretical approach to the use of cyber-infrastructure in geographical analysis', *International Journal of Geographical Information Science*, 23 (2): 169–93.

Wang, S., Anselin, L., Bhaduri, B., Crosby, C., Goodchild, M.F., Liu, Y. and Nyerges, T.L. (2013) 'CyberGIS software: a synthetic review and integration roadmap', *International Journal of Geographical Information Science*, 27 (11): 2122–45.

Weiner, D. and Harris, T.M. (2008) 'Participatory Geographic Information Systems', in J.P. Wilson and A.S. Fotheringham (eds), *The Handbook of Geographic Information Science*. Oxford: Blackwell Publishing. pp. 466–80.

Wilson, M.W. (2012) 'Location-based services, conspicuous mobility, and the location-aware future', *Geoforum*, 43 (6): 1266–75.

Wilson, M.W. and Stephens, M. (2015) 'GIS as media?', in S. Mains, J. Cupples and C. Lukinbeal (eds), *Mediated Geographies/Geographies of Media*. Dordrecht: Springer. pp. 209–33.

Wood, D. (1992) *The Power of Maps*. New York: Guilford.

3
DIGITAL MAPPING

JEREMY W. CRAMPTON

Digital mapping concerns the art and science of using digital technologies to deal with geospatial data. I say 'deal with' to include digitally mediated processes of collecting data, transforming them, weeding them out and combining them with other geospatial data. And beyond that, sharing and passing forward maps or mappable digital spatial data. Digital mapping may involve the production of maps, whether on a computer screen or displayed on mobile devices – although they may or may not be the ultimate product.

A problem with this sort of definition is that digital mapping is not a historically consistent object. Any definition I might contrive to fit today's digitally mediated landscape is not going to fit yesterday's or tomorrow's very well because the theory, technologies and praxes underpinning digital mapping have been and continue to be in flux. Even the term has changed over time. In the 1980s the terms 'computer-assisted cartography' or 'computer mapping' were used to refer to similar ventures. At that time, Mark Monmonier could confidently identify a 'technological transition' in cartography, and predict a near future where 'digital maps will displace the paper map' (Monmonier, 1982: 4). Looking back further, candidates for the 'first' digital map might include the work of John von Neumann at Princeton, who used the ENIAC computer after World War II to create the first map out of computer-generated data (Hall, 1992). Before that, Bletchley Park's Colossus computer was the first programmable digital computer (1941), and was used to decrypt messages from German High Command regarding the location of U-boats, which were then plotted on a large map.

But these candidates raise more questions than they answer. A search for origins, although it may give us a working timeline, should not obscure more interesting questions of what problems the techniques were meant to address, what forms of knowledge were created, nor indeed the socio-political context of their formation. Nevertheless, it does help us to understand both the history of our present situation and that it has not remained static.

A productive way into digital mapping is to consider it 'spatial media', meaning that geospatial information can be produced, shared and analysed to create value and to act as media for communicating other information (see Chapter 1). The key word here is 'shared'. Digital mapping, especially practised online as part of the geoweb, is a vital part of such 'new spatial media' (Crampton, 2010; Leszczynski, 2015). As spatial media, digital mapping is best understood as socio-technology. In other words, what is at stake are not only questions of technology, but also 'sites of potential relations between individuals; persons and places; and people, technology, and space/place' (Leszczynski, 2015). Spatial media have economic, political and, increasingly, legal ramifications (see Chapter 15).

In apprehending such a collection of issues, some authors interpret digital mapping and spatial media as comprising an assemblage (Dittmer, 2014). This is a complex term, but for our purposes we can use the definition of assemblage as 'a multiplicity which is made up of many heterogeneous terms and which establishes liaisons, relations between them … [i]t is never filiations which are important but alliances, alloys' (Deleuze and Parnet, 1987: 69). In other words, how do a wide variety of actors, institutions and knowledges form and reform and what work do they do in the world?

Two useful ideas can be drawn from assemblage theory. First, what is at stake is defined not only by what's going on within the assemblage, but also by what are known as its 'relations of exteriority' (DeLanda, 2006: 10). We have to look at how our subject (here, digital mapping) is nested and interrelated with other issues. For example, your new Apple Watch may be great at providing you with directions, but what rights does the government have to use it to track your location? To answer this question would mean enrolling legal and regulatory practices, rulings and knowledges. Second, assemblage theory points to capacities rather than properties of component parts. So for example, it is not so much the properties of spatial data we are interested in, as what work they do in the world. Or to put it another way, the relations of the properties to other properties. This has given rise to network theory and the Semantic Web (how meaning arises from these relations) (see Chapter 13). Note that while spatial properties are relatively computationally tractable in a geospatial database (think of a spreadsheet where each row is location and each column a certain property), data capacities are less tractable unless they too can be made calculable – hence the recent burgeoning interest in algorithms as ways to understand how capacities can be made calculable (Amoore, 2011; Kitchin, 2014).

Spatial media also allow us to understand the fact that today digital mapping is not necessarily a desktop-based process. Traditional 'Big GIS' (geographic information systems) such as Environmental Systems Research Institute (Esri), which has $1.4 billion in annual sales (Helft, 2015), is still going strong (although see my comments on 'zombie GIS' below), but the landscape is more diversified in mapping options (see Chapter 2). Online mapping, sometimes known as the geoweb (see Chapter 1), offers many alternatives, some of which are discussed more fully below.

CASE STUDIES

Sometimes the best way to get a feel for a topic is to look at specific cases. If digital mapping is self-evidently tied to the development and exponential increase in computing power (known as Moore's Law), and despite important precursors such as J.K. Wright's categorisation of spatial data, which was later adopted in GIS (Wright, 1944), a significant date in its recent development is 2005. In August of that year Hurricane Katrina struck the New Orleans area, and as devastated residents and others around the world sought updates on where the flooding had hit, they turned to a new online mapping service, Google Earth. Google Earth was launched in June 2005, and because of its ability to share views in small text files called keyhole markup language (KML), it quickly came to be used by the US government to distribute its aerial imagery. This made the virtual globe software very popular.

Google Earth was originally created by a company called Keyhole (named after the US spy satellite programme), which Google had bought in 2001 – hence keyhole markup language (KML). Keyhole was also the name of the company that originally developed the virtual globe, then known as Earth Viewer, and it can be better understood as an assemblage. For instance, Keyhole received start-up funding in early 2003 from In-Q-Tel, a venture capital company funded by the Central Intelligence Agency (CIA), and was almost immediately used in the Iraq war by the National Geospatial-Intelligence Agency (NGA, then known as NIMA). In 2008 I interviewed Avi Bar-Zeev, a co-founder of Keyhole (Crampton, 2008). He pointed out that in addition to funding, Earth Viewer required technology (compressed imagery, sufficient internet bandwidth, a decent user interface/user experience (UI/UX)), expertise (software engineers, a CEO with a background in satellite imagery and links to the US intelligence community) and the right socio-political-economic situation to be successful (the dotcom boom of the late 1990s). There are also unintended consequences, especially around privacy in StreetView (Google originally did not mask people's faces, and was sued and fined for privacy violations and now censors images). In fact, the invention of the technology has pushed the law to new interpretations of geolocational privacy too, especially when combined with other technologies such as the proliferation of commercial drones (small unmanned aerial vehicles or UAVs) (see Chapter 22). (Bar-Zeev later became Senior Manager at Amazon's Prime Air drone delivery service.)

Google Maps was also launched in 2005 and achieved success quickly (it became the largest online mapping company by 2009), again largely because Google wisely decided to 'open' their data through an application programming interface (API). The API allowed people to send a request using a special key for a Google basemap to sit under their own data. In this way, hundreds of thousands of map mashups were developed without the need for specialised software. The mashups lived on web pages and adopted a slippy map style (zoomable and interactive).

A second significant case study is the rise of OpenStreetMap (OSM). OSM is the Wikipedia of the mapping world, and exemplifies spatial media. OSM maps are often used as basemaps in other mapping products, such as ArcGIS, QGIS, Carto and Mapbox. OSM was conceived in August 2004 by Steve Coast, who has admitted that he did not quite know where it would take him. The map, which is now global in scope and clocks in at nearly 50 gigabytes, is contributed to by over two million mappers using GPS units to trace paths. Users can also access digital imagery at the site and trace out buildings, parks, lakes and so on (there is an extensive list of object types). A more recent alternative is to collect imagery by drone, correct the image for distortions (perspective) and add the imagery to OSM as a reference for digitising. A useful tool to do imagery georeferencing is MapKnitter (http://mapknitter.org), offered by Public Labs, which no digital mapper should do without. MapKnitter corrects the image and can export the result as a georeferenced tagged image file format (TIFF) image (geotiff). OSM data are free to use by anyone, for any purpose. If you wish you can change the data by updating them – directly editing the database, or by downloading and modifying them locally. You could even collect new imagery (e.g. by drone), georeference it with free tools such as MapKnitter and add it to OSM.

This process is a significant difference between OSM and Google Maps, to which it is much harder to submit updates after getting some rather controversial entries in a practice known as 'mapjacking' (including an outline of the Android logo urinating on the Apple logo). In this way then OSM may actually have better global coverage than Google Maps (especially in urban areas). Additionally, running and cycling mobile apps such as Strava, which have masses of user data, are a natural fit for importing into OSM (although for reasons of data quality this will probably not be automated without quality controls). On the other hand, Google will continue as a major digital mapping company. Its two-dozen or so self-driving cars (which already drive autonomously on real highways in four US states) mean that it will need to compete in map quality (as well as lasers, radar and sensors); not to mention it has one massive dataset that OSM does not, that is, StreetView. Competitors such as Uber are also focusing on comprehensive mapping databases, and in 2015, an agglomeration of German auto manufacturers, BMW, Daimler and Audi, purchased Nokia's 'HERE' map business (previously known as Navteq) for $3.1 billion. The business of high-precision maps is obviously a significant one as self-driving vehicles become more prevalent.

WHO IS INVOLVED?

Prior to the advent of the personal computer in the 1980s and the widespread adoption of graphic user interfaces (GUI), digital mapping was the province of an elite few. As Monmonier (1982) hinted, digital mapping has come to replace paper mapping to such an extent that the latter has gained a modicum of 'artisanal' quality. (3D printing has also helped.) In one of the earliest textbooks on computer mapping, Monmonier cited over 200 available mapping packages as of 1977. These included those devised by government agencies such as the United States Geological Survey (USGS) and the Census Bureau (which had developed its DIME files of all US streets by the 1970 census), and state governments such as Minnesota, which created a GIS Land Management Information Center in 1977. Harvard's Laboratory for Computer Graphics and Spatial Analysis, a major research lab founded in 1965 by Howard Fisher, also developed software including SYMAP and ODYSSEY (which reputedly formed the basis for Esri's own products). Subsequent presidents of three GIS companies worked at the lab (Jack Dangermond – Esri; Howard Slavin – Caliper; and Lawrie Jordan – ERDAS) (Chrisman, 2006).

By the early 1980s, Dobson (1983) could claim that computers integrated all forms of geographic inquiry. Automation (as he saw it) was 'essentially neutral'

(Dobson, 1983: 135) and would make both the 'humanist' and the scientist more productive through use of new tools. The concern was to integrate mapping, then seen as largely a map display capability, with analysis, or spatial problem solving. (The same concern had led the Harvard Lab to add 'Spatial Analysis' to its original name.) The map by itself was insufficient. In a commentary on Dobson, Cowen (1983) suggested 'that maps as ends unto themselves are of little value to decision-makers' (Cowen, 1983: 339) and this was why an interesting government mapping initiative, the Decision Information Display System (DIDS) failed – it was merely a 'magical map making machine' (1983: 339).

Interestingly, DIDS, which was developed by the National Aeronautics and Space Administration (NASA) for the Carter White House and aimed at integrating data across government agencies for display 'onto a screen for viewing by many people in a large room' (Monmonier, 1982: 146), bears some resemblance in principle to 'Project Cybersyn' (Medina, 2011). Cybersyn was conceived for the Allende government in Chile in the early 1970s by Stafford Beer, a British cybernetician, but was never operational. Although it was not a mapping system per se, Beer envisioned it along the same lines as the Churchill government's 'vast map in the war-time Operations Room' (Beer quoted in Medina, 2011: 34). From a central control room, Chilean officials and everyday workers would be able to access flows of information about the Chilean economy and the public mood about policies – all in real time across a computer network (not the internet, which was barely a year old in 1970, but a series of repurposed Telex machines). Medina (2011: 6) also offers a useful corrective to Dobson's claim of technological neutrality: 'technologies are not value-neutral', she argues, 'but rather are a product of the historical contexts in which they are made'. Both DIDS and Project Cybersyn exemplify some of the grandeur of big data, and no doubt some of its hubris. They also exemplify the need to understand developments in terms of a genealogy rather than a history of firsts: Beer was inspired by Churchill's war-room map, which in turn was inspired by strategic military maps going back centuries. Cybersyn was related to space Mission Control operations rooms, and today's smart cities (see Chapter 19).

Today, Big GIS (e.g. ArcGIS, DIDS, and enterprise GIS) have been supplemented (some would say replaced) by other options. The digital mapping landscape is incredibly diverse and richly functional. While this is partially a question of technology (especially the personal computer, the availability to the public of good GPS since 2000, online mapping technologies, and the advent of the smartphone), it is also a question of philosophy. If Project Cybersyn was meant to give the ordinary worker in Chile a bigger say in the means of production, then it prefigured what would only come later for most, that is, the means of map production for oneself. Goodchild has identified this shift in thinking as 'Volunteered Geographic Information' (VGI, see Chapter 12) and if that term is known mostly to academics, its principles are much more widely practised. VGI inverts the traditional model of mapping where a skilled cartographer collates and processes data, accesses specialised production tools, designs a single map and passes it on to a client. Today, a typical workflow is that an individual collects geolocational data with a smartphone app such as Strava or Fulcrumapp, pools the data with that of workmates, exports it using modern geospatial file formats such as GeoJSON, and imports it into geoweb environments such as Carto or Mapbox. Both these latter two companies are new. CartoDB was founded in Spain in 2007 by an agricultural engineer and a computer scientist for the purposes of better data visualisation (Solana, 2015). Mapbox was founded in 2010 and has made a huge impact on web-based digital mapping, hosting and visualising spatial data, but more importantly allowing developers to design new maps and promoting the cross-platform mapping stylesheet called CartoCSS. Both companies work well with tiling (Mapbox Studio allows custom vector tiles) and OSM and are web-native.

Another option to Big GIS might be to use desktop-based QGIS, an open-source extensible GIS that competes with ArcGIS. Under this scenario, the fears of Dobson and Cowen that the map is insufficient are revealed to misunderstand that the map itself is analysable and provides actionable intelligence. This last point is especially noticeable in the context of the smart city. As we have seen, the rise of the smart-city approach to planning over the past two decades (the term seems to have originated in the mid-1990s), the use of maps and other indicators in what are sometimes called 'urban dashboards' in operations centres has also underlined how digital maps can aid decision-making (Mattern, 2015; see also Chapter 7). Here maps do not stand alone, but are part of an ever-forming assemblage. As Kitchin et al. (2015) point out, dashboards (and by extension digital maps) are 'co-produced' with the wider institutional landscape, and are affected by and in turn affect the city. Here, maps could not be anything other than spatial media.

WHAT IS NECESSARY TO UNDERSTAND ABOUT DIGITAL MAPPING?

In the previous section we discussed some of the significant players and events to do with digital mapping. But how should we understand these developments? In this section I identify three key factors for understanding digital mapping as spatial media: (1) 'old school' Big GIS with black-boxed proprietary software is less dominant; (2) the information and sharing economy means that the map is not the end product, but rather part of a whole series of geospatial services and content that are shared and networked; (3) digital mapping is an assemblage and the individual mapper working alone with bespoke technologies is giving way to the networked collaborator. I discuss these briefly below.

Is GIS dead? Well, yes and no! It was not too long ago that cartographers fretted that GIS would kill cartography, but today companies like Carto and Mapbox, and web-based mapping services such as MapKnitter, mean that cartography (as the art and science of making maps) is resurgent. And I have already mentioned the new markets that both UAVs and self-driving cars are opening. Big GIS – exemplified by Arc/Info and ArcGIS – may be 'dead', but like a zombie, still stumble around out of habit. There are a lot of companies using ArcGIS: over a third of a million according to Esri. GIS is changing to the extent that we can no longer call it 'GIS'. But Esri is a smart company, and has started to make the turn towards the cloud, towards online, searchable data sharing (i.e. not just map sharing) and to 'story maps' or curated maps, images, text and video that tell a meaningful narrative (very useful in journalism, for example). But it will take a while for the supertanker to change course. In the meantime those who need maps but not Big GIS can use lots of different 'little GISs' (for a very useful review of 35 options, see Roth et al., 2015).

Second, the de-emphasis on the map as end product. This has been mentioned several times throughout this chapter, but consider OSM. It is not so much 'a' map but rather a symbolised, ongoing geospatial database that can be viewed, extracted, modified, added as a basemap, and used for navigation or for humanitarian relief efforts (Haiti and Nepal are examples where OSM quickly provided detailed maps through concerted community mapping efforts).

Third, digital mapping comprises an assemblage. Digital mapping as spatial media is by definition a social activity. In the 1960s, cartographers such as Arthur Robinson developed a model of cartography as a map communication model (MCM). The impetus for this was very reasonable, namely that mapping was moving from being a craft industry with highly artistically talented individuals such as Erwin Raisz, Armin Lobeck and Richard Edes Harrison, to something that needed more formal rules of study. The MCM outlined a process whereby an expert mapmaker could produce maps on the production line method, drawing on rules and accepted practices rather than individual artistic talent. Like a factory production line, however, this model centralises production and responds to demand from the marketplace. With the opening-up of mapping by APIs, map mashups, VGI and geoweb mapping companies, this process becomes more networked and shared. It does not mean the end of mapping expertise but it does represent a huge shift (e.g. more towards coding). And the sharing is not at the end of the process but throughout, so that, for example, even primary data collection can be crowdsourced. So if the MCM is Fordism, then today's digital mapping is 'just in time' production or even DIY maker-spaces.

UNDERSTANDING MAPPING IN THE AGE OF THE NEOLIBERAL ALGORITHM

There are two things that appear as if they are going to be important to the future of digital mapping. The first is that while it is easy to say digital mapping comprises an assemblage, we do not know the full implications of that yet. What are its relations to big data? How does digital mapping play a role in the internet of things or the smart city? What is the relative importance of mapping per se to geolocational information more generally for issues such as geoprivacy (see Chapter 22)? How will legal determinations play out with regard to what is legally 'reasonable' geolocational tracking? Given that individuals 'shed' copious amounts of geolocational information in the course of their daily activities, and that it is upon these algorithmic 'data derivatives' that decisions are made (Amoore, 2011), then how can our privacy be assured? Is it a question of 'rights' (even if rights for some means lack of rights for others) such as the (problematic) right to be forgotten? Will countries such as the USA and the UK rein in bulk or so-called dragnet surveillance in favour of more targeted, warrant-driven investigations?

Second, how is digital mapping being reconfigured in the age of neoliberalism? By this I mean the processes of making everything calculable and monetisable, and the way new economic zones are being colonised for value extraction. An example of the latter – in what might actually be the first new space colonised since the scramble for Africa in the 1880s, is vertical airspace and the advent of the commercial drone. Perhaps 600,000 people are in the air at any one time, but this still leaves most of the Earth's volume (1080 billion km³ according to NASA, compared to 1.4 km³ billion for the volume of Earth's oceans) available for occupation.

Small companies making unmanned aerial vehicles are not so small any longer (the world's biggest is DJI, which in 2015 sought a $10 billion valuation). DJI, Google[x] Project Wing and Amazon's Prime Air are members of a coalition pushing for more business-friendly regulations in the USA (which currently prohibit flying drones beyond line of sight). If the Federal Aviation Administration (FAA) were to relax its prohibitions on commercial drones, how would geospatial information play a role? More generally, how are the various apps we have in smartphones extracting value from our geolocational data?

To answer these questions we need more, and better, critical histories of mapping as a socio-technological development. Kitchin (2014) has usefully outlined six approaches for thinking critically about geolocational algorithms. In particular he urges that we study how the objects of our research do work in the world (assemblage theory's emphasis on capacities) and the wider socio-technological contexts. These are also useful for researching digital mapping.

REFERENCES

Amoore, L. (2011) 'Data derivatives: on the emergence of a security risk calculus for our times', *Theory, Culture & Society*, 28: 24–43.

Chrisman, N. (2006) *Charting the Unknown: How Computer Mapping at Harvard Became GIS*. Redlands, CA: Esri Press.

Cowen, D.J. (1983) 'Commentaries on "automated geography": automated geography and the DIDS (Decision Information Display System) experiment', *The Professional Geographer*, 35: 339–40.

Crampton, J.W. (2008) 'Keyhole, Google Earth, and 3D Worlds: an interview with Avi Bar-Zeev', *Cartographica*, 43: 85–93.

Crampton, J.W. (2010) *Mapping: A Critical Introduction to Cartography and GIS*. Oxford: Wiley-Blackwell.

DeLanda, M. (2006) *A New Philosophy of Society: Assemblage Theory and Social Complexity*. London: Continuum.

Deleuze, G. and Parnet, C. (1987) *Dialogues*. New York: Columbia University Press.

Dittmer, J. (2014) 'Geopolitical assemblages and complexity', *Progress in Human Geography*, 38: 385–401.

Dobson, J.E. (1983) 'Automated geography', *The Professional Geographer*, 35: 135–143.

Hall, S.S. (1992) *Mapping the Next Millennium: The Discovery of New Geographies*. New York: Random House.

Helft, M. (2015) 'You can't kill Jack Dangermond's company. Try, and it will only get stronger', *Forbes*. Available at: http://www.forbes.com/sites/miguelhelft/2015/03/31/you-cant-kill-jack-dangermonds-company-try-and-it-will-only-get-stronger/ (accessed 21 July 2016).

Kitchin, R. (2014) 'Thinking critically about and researching algorithms', *Information, Communication and Society*. DOI: 10.1080/1369118X.2016.1154087

Kitchin, R., Maalsen, S. and McArdle, G. (2015) 'The praxis and politics of building urban dashboards', *Programmable City Working Paper 11*. Available at: http://ssrn.com/abstract=2608988 (accessed 21 July 2016).

Leszczynski, A. (2015) 'Spatial media/tion', *Progress in Human Geography*, 39 (6): 729–51.

Mattern S. (2015) 'Mission control: a history of the urban dashboard', *Places Journal*. Available at: https://placesjournal.org/article/mission-control-a-history-of-the-urban-dashboard/ (accessed 21 July 2016).

Medina, E. (2011) *Cybernetic Revolutionaries: Technology and Politics in Allende's Chile*. Cambridge, MA: MIT Press.

Monmonier, M. (1982) *Computer-Assisted Cartography: Principles and Prospects*. Englewood Cliffs, NJ: Prentice-Hall.

Roth, R.E., Donohue, R.G., Sack, C.M., Wallace, T.R. and Buckingham, T.M. (2015) 'A process for keeping pace with evolving web mapping technologies', *Cartographic Perspectives*, 78: 25–52.

Solana, A. (2015) 'CartoDB: from Alcatraz to elephants, startup puts open source visualisation on the map', *ZDNet*. Available at: http://www.zdnet.com/article/cartodb-from-alcatraz-to-elephants-startup-puts-open-source-visualisation-on-the-map/ (accessed 21 May 2016).

Wright, J.K. (1944) 'A proposed atlas of diseases', *Geographical Review*, 34: 642–52.

4
DIGITALLY AUGMENTED GEOGRAPHIES

MARK GRAHAM

The term 'spatial media' has traditionally been used to describe the intersections between information and geography. It signifies information that describes, or is about, a particular place. A street map of Chicago, geographic data files about Copenhagen, a postcard with a picture of Oxford on it, a travel guide to Sweden: are all examples of spatial media; in other words, information about geography.

It was only relatively recently that geographic information became so easily infused into spatial media. For most of human history, geographic information was tethered onto particular parts of the world. It passed from person to person, often changing because it was so difficult to attach it to stable containers. But then, a succession of technological advancements (like papyrus and the printing press) allowed for the invention of books, newspapers, maps, and other media. What these mediums had in common was that they fixed geographic information to its containers: they made it immutable. A paper map, for instance, could be moved from place to place without the map itself changing.

This chapter, however, is about the recent, dramatic transformation of spatial media. What has occurred is not just a move from mutable geographic information to immutable geographic information, but also the increasing proliferation of what could be called 'augmented spatial media'. Instead of solely being fixed to containers, information can now augment and be tethered to places; it can form parts of the layers or palimpsests of place (Graham et al., 2013). A building or a street can now be more than stone, bricks and glass; it is also constructed of information that 'hovers over' that place, invisible to the naked eye, but accessible with appropriate technological affordances. In other words, while it has long been argued that 'the map is not the territory' (Korzybski, 1948; Harley, 1989; Crampton, 2001), this chapter argues that the map is indeed becoming part of the territory.

HISTORIES OF AUGMENTED MEDIA

Historically, augmented media has fallen into two categories: *information about places*, and *information that can be accessed in places*. The difference between the latter and the former is that the spatial media containing information about a place are actually used or deployed while in the place or navigating through the place. It is worth remembering that both types of spatial media have existed for a long time. You could go to a library to access reference material about the road grid of a town, or you could take a map with you to town and access information about the place while you are actually there.

What is changed is that we have developed technologies to effectively separate content from its containers. Previously information was physically stuck onto its containers: words on a page, lines on a map, pictures on a postcard. To move information, you had to move the container of information. But as ever more information is digital and digitised, the costs associated with moving geographic information recede towards zero.

Early instances of this include online directory information about places (e.g. the Yahoo! directory), online-maps data (e.g. Mapquest) and a range of user-generated content about specific places (e.g. 'placemarks' and .kml files). More recent examples include encyclopaedic information (e.g. Wikipedia or Baidu Baike articles about a place), geotagged photographs (on platforms like Flickr or Instagram), social media content (e.g. tweets tagged to a place), explicitly place-based digital social platforms (e.g. Facebook's Places or Foursquare) and review-based platforms (e.g. Tripadvisor or Google's map-based reviews).

In all of those examples, information is simultaneously fixed to a particular part of the world while it is unfixed from its containers. It can then be accessed on mobile devices like mobile phones, digital watches and other portable computing devices. A review of a restaurant called Rusholme Chippy in Manchester submitted through Google Maps is constantly linked to that place: it is visible if a user specifically searches for that restaurant, if they are doing a search for 'fish and chips restaurants', 'kebab shops' or even just 'restaurants' in Manchester, or if they are accessing geographic content about the surrounding area and click on the digital representation of that restaurant. In all of those cases, that content has been assigned a location, and anybody using specific hardware/software/platform combinations who navigates to that location can access the relevant information.

This is ever more the case with the move towards digital architectures based on linked data and the Semantic Web (see Chapter 13). The purposes of both linked data and the Semantic Web are to create frameworks allowing data to be easily shared and used across sites and platforms. Said differently, not only has content become untethered from its physical containers: it is becoming untethered from its digital ones as well (Ford and Graham, 2015). For instance, a piece of information like the name of the capital city of Rwanda (Kigali) will no longer just be stored as an unstructured piece of information in the Wikipedia pages about Rwanda, Kigali, East Africa, capital cities, and so on; that content is increasingly being stored in, and accessed from, linked datasets (like Wikidata), and from there being used to populate relevant services that need the information (like Wikipedia, and Google searches).

What this means is that with the requisite digital infrastructure, the idea of spatial media can refer to more than just space being the subject of media. Content is becoming unfixed from its containers, but re-fixed to locations. Ever more information is augmenting places, and ever more places are becoming augmented. Therefore, as our world becomes increasingly augmented and augmentations become part of the cities that we live in, it is crucial not just to recognise the existence of augmented media, but also to ask who controls that information, who has the ability to access it, and who has the power to change it.

KEY ISSUES

There are four key issues that we should concern ourselves with when considering how informational augmentations impact everyday life: access, control, representation and participation.

Access: If important parts of places are digital, an important first consideration is who it is that can actually access those augmentations. It is worth remembering that a majority of the world's population has still never used the internet. Even in countries with some of the highest internet penetration rates, significant numbers of people are left out. In the UK, for instance, non-users make up about one-fifth of the population and tend to be poorer, older and live in more economically deprived parts of the country (like the north-east) (Dutton and Blank, 2013; Blank et al., 2015). Furthermore, access is not a simple binary (cf. Graham and Dutton, 2014). Of the three billion people who use the internet, many are forced to use it in very restricted ways because of economic necessity (e.g. metered plans, bandwidth caps), technical limitations (e.g. slow connectivity speeds) and government restrictions (e.g. censorship or surveillance).

Finally, it is worth noting that the world's largest companies who control key platforms need to augment the world with digital content (Google, Microsoft and Facebook) are investing heavily in projects that aim to bring access to the world's currently disconnected. Facebook, for instance, is investing in the Internet.org project that offers free internet access to mobile phone owners in eight countries across the Global South (as of May 2015). However, the connectivity offered is severely limited and restricts users to accessing only a few platforms (Facebook being one of them). Such a strategy ensures that an ever-growing user-base reinforces the control over key sites of digital augmentations that large entities, like Facebook, already possess.

Control: Although the internet, in theory, allows for an enormous diversity of digitally mediated practice, use tends to cluster around a few central sites. People tend not to want to use multiple platforms or sites to access information. The net result then is that only a handful of entities control the vast majority of people's interactions with augmented content (Fuchs, 2014). Facebook and Twitter, for instance, mediate a huge number of place-based social and information-sharing practices; Wikipedia is the definitive site for sharing encyclopaedic content; and Google mediates an enormous amount (and proportion) of the world's searches for local information.[1] With the notable exception of Wikipedia, these mediators are all for-profit companies.

Representation: There are two primary mechanisms in which particular parts of the world become represented in augmented media. First, one of the companies mentioned above will build a spatial dataset (which is usually assembled from a range of other already-existing datasets) and take editorial control over how places should be represented. Second, a platform can rely on user-generated content to form the information that is to be augmented. Here, the typical model is that everyday users generate the bulk of content and participate in key editorial decisions.

Yet both models produce highly uneven outcomes. Google's spatial database, for instance, contains much more content about places in Europe and North America

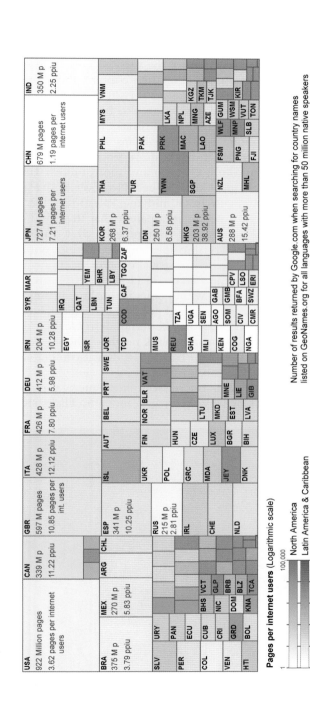

FIGURE 4.1 Number of Google search results for country names

Source: De Sabbata and Graham – http://geography.oii.ox.ac.uk/?page=geographies-of-google-search

than it does about places in Africa or South America. This is not because there are more people or more cities in the former than in the latter (see Figure 4.1).

These uneven geographies of content are also manifested in the ways that different types of content can augment the same parts of the world. In research that was conducted to look at the relative amount of augmented content about the same places in different languages (see Graham and Zook, 2011, 2013, for more details), it was found that traditionally marginalised places can also become marginalised in digital representations. Figure 4.2, for instance, illustrates the relative amount of English and French-language content in Google Maps about different parts of Eastern Canada. As might be expected, there is more English-language content about predominantly English-speaking Ontario and more French-language content about predominantly French-speaking Quebec. However, if the geolinguistic contours of augmented content in Israel and the Palestinian Territories are similarly mapped (see Figure 4.3), it becomes apparent that there is a denser cloud of Hebrew content than there is of Arabic content, even over places under the political control of a predominantly Arabic-speaking population. Within the Palestinian Territories, Arabic-language searches for spatial content yield only 5–15% of the number of search results that the same searches in Hebrew do. This is not just the case at the national or regional scale; even within cities inequalities in augmented content that mirror underlying social, economic and political inequalities have been documented. For instance, predominantly white neighbourhoods in New Orleans are layered with more information than predominantly black neighbourhoods (Crutcher and Zook, 2009), and Palestinian East Jerusalem has far less digital content about it than Israeli-controlled West Jerusalem (Carraro, 2015).

Relying on user-generated content does not necessarily alter some of these uneven geographies of augmented content. Wikipedia, Wikivoyage, Panoramio, Flickr and every

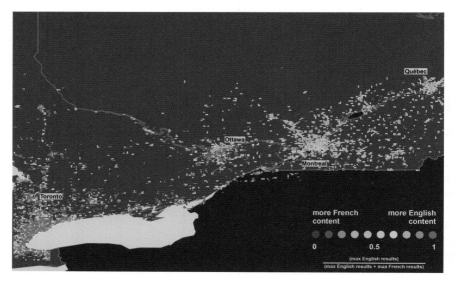

FIGURE 4.2 Ratio of French to English content in Ontario and Quebec, Canada

Source: Graham and Zook (2013)

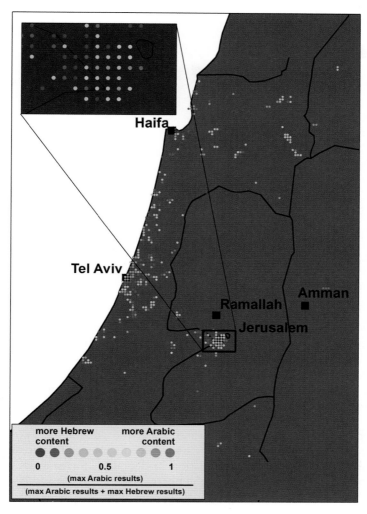

FIGURE 4.3 Ratio of Hebrew to Arabic content in Israel and the Palestinian Territories

Source: Graham and Zook (2013)

other platform that aggregates user-generated geospatial content are similarly characterised by concentrations of content in the Global North and only small amounts of
information produced about the Global South (cf. Graham et al., 2011, 2014). For
instance, there is more content in Wikipedia about the Netherlands than all of Africa
combined. Yet, more is at stake than just the under-representation of some places. There
are countless examples of buildings, monuments, streets, parks, cities, lakes, borders and
countries that are highly contested by different groups of people. It thus follows that
the digital spatial content that augments them will also be contested. But in platforms
like Wikipedia that rely on consensus, it is rarely minority voices that win arguments
(see Graham, 2013, for an example of how Palestinian and Israeli perspectives about
how Jerusalem should be spatially represented in Wikipedia played out in practice).

average number of edits to Wikipedia

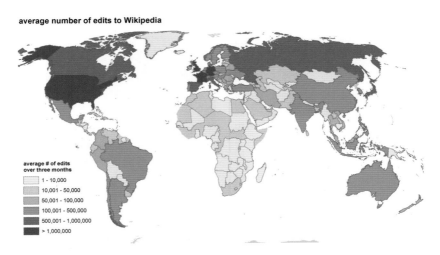

FIGURE 4.4   Edits to Wikipedia

Source: Graham (2014)

Participation: Not only is the actual content that augments parts of our world skewed and uneven, but so too are the patterns of participation that end up creating it. For instance, the vast majority of information created in Wikipedia is created by people in the Global North (despite most users of Wikipedia being in the Global South) (see Figure 4.4). In Hong Kong alone, there are more edits to Wikipedia than the whole continent of Africa combined.

Research by Hecht and Gergle (2010) and Graham et al. (2014) illustrates that content tends to have a local bias, meaning that the geographically concentrated nature of edits can further reinforce the uneven geographies of augmented content described above. Furthermore, participation in digital platforms tends to massively exclude women and ethnic minorities (Stephens, 2013; Crutcher and Zook, 2009). It is thought that fewer than 15% of Wikipedia's editors are women (Lam et al., 2011).

AUGMENTED MEDIA OF THE CITY

The previous sections have made three key arguments. First, contemporary spatial media are able to easily separate content from its containers and attach it to places. This means that places are increasingly augmented with digital content that is as much a part of the place as bricks or buildings are. In other words, spatial representations are no longer just mediations of information about place; they are integrally a part of place.[2] Second, much of the digital content that augments places is controlled (and owned) by a small number of for-profit companies (with the notable exception of Wikipedia). Third, the content that exists is highly uneven, with practices of participation and geographies of representation highly skewed towards a small number of people, places and perspectives.

If we accept the first argument (that digital spatial representations now form part of places, rather than being just mediations of them) and we find the implications of the second and third arguments problematic, then it is worth paying closer attention to a few further issues that arise when the city itself is mediated.

As an increasing amount of everyday geographic life becomes augmented and technologically mediated, it is worth reflecting on the censorship that can occur in the platforms, systems and databases mentioned earlier in this chapter. Some censorship is explicitly performed by governments that seek to ensure that places are augmented by only certain kinds of information. The Chinese government, for instance, goes to great lengths to limit the kinds of information available through digital platforms (Zittrain, 2008). It is no accident that searches for information about Tiananmen Square using an American-based or China-based service yield radically different results (see Figure 4.5).[3]

The kinds of curation mentioned above are carried out overtly by actors who want to ensure that certain kinds of information exist, or do not exist, about different parts of the world. Yet an increasingly prevalent way in which spatial information is filtered is through algorithms and code (cf. Kitchin and Dodge, 2011). This algo-rithmic filtering reveals some information about places to users and conceals other content. But algorithms are rarely designed to augment places with random infor-mation. Instead, their designers imbue them with specific goals, often to give people contextually relevant information, and to make commercially aware and sensitive choices. But those goals can serve to reinforce spatial filter bubbles (Graham and Zook, 2013; Pariser, 2011). These geographic filter bubbles can be pernicious in the ways that they not only strip away serendipity from spatial encounters, but actually reinforce digital ghettoisation in which different groups of people are directed to very different parts of their urban environments. In Figure 4.6, for instance, Hebrew- and Arabic-speaking users of Google Maps are directed to fundamentally different parts of Tel Aviv.

FIGURE 4.5 Searches for Tiananmen Square using Google (US) and Baidu (China)

Source: http://blockedonweibo.tumblr.com/

The two concerns mentioned above can produce social, economic and political outcomes undesirable for citizens or residents of any augmented place. In both cases, important decisions are made about not just what should and should not be included and excluded, but how places should be represented. In the case of state censorship, the ability for (or constraint on) redress tends to be apparent to most citizens. Privately owned datasets and platforms are relatively unaccountable to citizens, but they represent a visible entity that can be petitioned and challenged in instances when augmented content might be unfair or unrepresentative.

User-generated content, in contrast, by virtue of its openness, is more inclusive and participatory. But that decentralised nature also renders it even more difficult to challenge dominant voices. When Google, for instance, represents a city in a particular

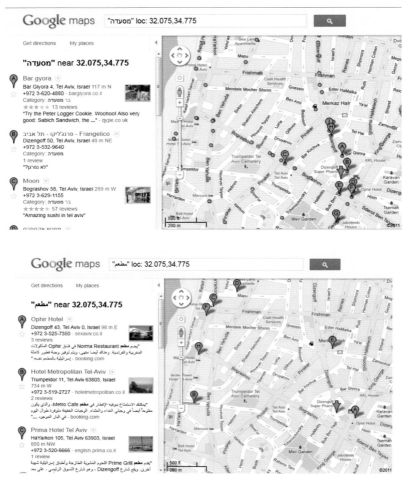

FIGURE 4.6   Searches in Google Maps for a restaurant in Hebrew and Arabic from the same starting point in Tel Aviv

Source: Graham and Zook (2013)

way, most users of that information know to blame Google if they disagree with it. In contrast, the mechanisms for altering content on open systems like Wikipedia can paradoxically be more opaque when people are unfamiliar with whom it is they actually have to petition for change (Ford and Graham, 2015). An amorphous and decentralised network of editors can be hard to sway with minority perspectives.

As the state rolls back from its engagement with control over geographic information, ever more augmentations of place are controlled either by large corporate interests or the logic of peer production (Leszczynski, 2012). In either case, under this neoliberal political economy of augmented content, neither the market nor the crowd have a mandate to be democratic or inclusive.

FUTURE DIRECTIONS

It is unclear how future technological innovations, social practices and economic pressures will reshape how people engage with spatial augmentations. But it is unlikely that we will move back to a world in which digital augmentations of place are not prevalent. Therefore, at this nascent moment in the historical trajectory of augmented media, it is worth asking what future generations of researchers and citizens can be doing to better understand augmented media, and what they can be doing to be able to enact change in augmented mediations.

First, more transparency and knowledge are needed. Augmented media shape everyday life, but, with a few exceptions, both augmented content and mediations of that content are produced, regulated and represented in highly opaque ways. The first step to bringing into being any sort of improvement is to better understand contemporary information geographies. This is not as simple as mapping what is 'there'. Rather, filter bubbles, personalisation, the increasing separation of content from its containers due to the Semantic Web (see Chapter 13), and the unanticipated outcomes of automated algorithms mean that we need to develop better ways of asking older questions about participation and representation.

Second, more critical reflection into alternatives is needed. If we accept that augmented media are here to stay, what specific strategies can be enacted to counter some of the negative outcomes outlined in this chapter? At the moment, we have large organisations like Google controlling important facets of our towns and cities. This observation sits within a longer history of critique about the role that market-based mechanisms play in organising spatial everyday practice and consequently producing unjust outcomes (cf. Attoh, 2011). 'Peer production' is then often presented as a panacea that will allow more openness, a broader base of participation, and more democratic layers in our spatial environments. But, as this chapter has demonstrated, within peer-produced augmented content, significant bias persists. Neither the market nor the crowd are producing optimal outcomes. What then is the solution?

Perhaps there is a need for large-scale public-sector involvement, investment and regulation: public web indexes or revitalised mapping initiatives to counter what Leszczynski (2012) refers to as 'technoscientific capitalism'. Perhaps we can follow Lanier's (2013) vision of allowing users to monetise their own data, as a way of preventing the ever-increasing accumulation of power in the hands of a few private entities. Or perhaps what is also needed is a range of micro-political challenges to the digital status quo (Shaw and

Graham 2017): strategies to inappropriately appropriate platforms, misrepresent, over-represent, delete, amend and pervert information. This de-valuing of data as a means of protest can force attention onto the constellations of technology, information and social practice that easily slip into the background and otherwise become normalised.

In any case, augmented media are now a feature of billions of people's everyday lives. It has become increasingly easy to attach information to places. Furthermore, augmented information is itself always mediated: through machines, databases, people and code. It is that information and those mediations that we need to now ensure we better understand through more targeted questions, methods and data. Doing so allows us to better ask not just what the digital layers of our cities look like, what they include and who controls them, but also how we begin to change them.

ACKNOWLEDGEMENTS

This chapter draws on and reworks material from Graham, M., Zook, M. and Boulton, A. (2013) 'Augmented reality in the urban environment', *Transactions of the Institute of British Geographers*, 38 (3): 464–79; Graham, M. and Zook, M. (2013) 'Augmented realities and uneven geographies: exploring the geo-linguistic contours of the web', *Environment and Planning A*, 45 (1): 77–99; Graham, M. (2014) 'Internet geographies: data shadows and digital divisions of labour', in M. Graham and W.H. Dutton (eds), *Society and the Internet: How Networks of Information and Communication are Changing Our Lives*, Oxford: Oxford University Press, pp. 99–116; and Graham, M., De Sabbata, S. and Zook, M. (2015) 'Towards a study of information geographies:(im)mutable augmentations and a mapping of the geographies of information', *Geo: Geography and Environment* 2 (1). 88-105.

NOTES

1. There are two important caveats that should be noted here. First, none of this discussion really applies to the context of China, which has its own domestic alternatives (e.g. Baidu Baike and Weibo). Second, other important platforms exist that mediate more niche, or specialised, information about places, but, for the sake of simplicity, this chapter focuses on dominant and general purpose platforms.
2. But, importantly, they never cease to be mediated.
3. A more detailed overview of censored geographic information is available in the following Wikipedia article: http://en.wikipedia.org/wiki/Satellite_map_images_with_missing_or_unclear_data

REFERENCES

Attoh, K.A. (2011) 'What kind of right is the right to the city?', *Progress in Human Geography*, 35 (5): 669–85.

Blank, G., Graham, M. and Calvino, C. (2015) 'Local geographies of digital inequality', manuscript under review.

Carraro, V. (2015) 'Grassroots mapping: experiences from occupied Jerusalem', *Presentation at Digital Geographies*, Milton Keynes, 24 March 2015.

Crampton, J.W. (2001) 'Maps as social constructions: power, communication and visualization', *Progress in Human Geography*, 25 (2): 235–52.

Crutcher, M. and Zook, M. (2009) 'Placemarks and waterlines: racialized cyberscapes in post-Katrina Google Earth', *Geoforum*, 40 (4): 523–34.

Dutton, W.H. and Blank, G. (2013) *Cultures of the Internet: The Internet in Britain*. Oxford: Oxford Internet Survey 2013 Report.

Ford, H. and M. Graham (2015) 'Semantic cities: coded geopolitics and the rise of the semantic city', in R. Kitchin and S.Y. Perng (eds), *Code and the City*. London: Sage. pp. 200–14.

Fuchs, C. (2014) *Social Media: A Critical Introduction*. London: Sage.

Graham, M. (2013) 'The virtual dimension', in M. Acuto and W. Steele (eds), *Global City Challenges: Debating a Concept, Improving the Practice*. London: Palgrave. pp. 117–39.

Graham, M. (2014) 'Inequitable distributions in internet geographies: the Global South is gaining access but lags in local content', *Innovations*, 9 (3–4): 17–34.

Graham, M. and Dutton, W.H. (eds) (2014) *Society and the Internet: How Networks of Information and Communication are Changing Our Lives*. Oxford: Oxford University Press.

Graham, M. and Zook, M. (2011) 'Visualizing global cyberscapes: mapping user-generated placemarks', *Journal of Urban Technology*, 18 (1): 115–32.

Graham, M. and Zook, M. (2013) 'Augmented realities and uneven geographies: exploring the geo-linguistic contours of the web', *Environment and Planning A*, 45 (1): 77–99.

Graham, M., Hale, S.A. and Stephens, M. (2011) *Geographies of the World's Knowledge*. London: Convoco! Edition.

Graham, M., Zook, M. and Boulton, A. (2013) 'Augmented reality in the urban environment: contested content and the duplicity of code', *Transactions of the Institute of British Geographers*, 38 (3): 464–79.

Graham, M., Hogan, B., Straumann, R.K. and Medhat, A. (2014) 'Uneven geographies of user-generated information: patterns of increasing informational poverty', *Annals of the Association of American Geographers*, 104 (4): 746–64.

Harley, J.B. (1989) 'Deconstructing the map', *Cartographica*, 26 (2): 1–20.

Hecht, B. and Gergle, D. (2010) 'On the "localness" of user-generated content', in *CSCW '10: 2010 ACM Conference on Computer Supported Cooperative Work*, Savannah, GA. pp. 229–32.

Kitchin, R. and Dodge, M. (2011) *Code/Space: Software and Everyday Life*. Cambridge, MA: MIT Press.

Korzybski, A. (1948) *Science and Sanity: An Introduction to Non-Aristotelian Systems and General Semantics*, 3rd edition. Lakeville, CN: The International Non-Aristotelian Library Pub. Co.

Lam, S.K., Uduwage, A., Dong, Z., Sen, S., Musicant, D.R., Terveen, L. and Riedl, J. (2011) 'WP: Clubhouse? an exploration of Wikipedia's gender imbalance', in *Proceedings of the 7th International Symposium on Wikis and Open Collaboration, WikiSym '11*. New York: ACM. pp. 1–10.

Lanier, J. (2013) *Who Owns the Future?* New York: Simon & Schuster.

Leszczynski, A. (2012) 'Situating the geoweb in political economy', *Progress in Human Geography*, 36 (1): 72–89.

Pariser, E. (2011) *The Filter Bubble: What the Internet is Hiding from You*. London: Viking.

Stephens, M. (2013) 'Gender and the geoweb: divisions in the production of user-generated cartographic information', *GeoJournal*, 78 (6): 981–96.

Zittrain, J. (2008) *The Future of the Internet and How to Stop It*. New Haven, CT: Yale University Press.

5

LOCATIVE AND SOUSVEILLANT MEDIA

JIM THATCHER

According to Dan Evans of Cisco, the year 2008 marked a turning point in humanity's relationship to the internet – for the first time there were more devices connected than human beings currently alive (Evans, 2011). These devices are not simply phones, tablets and computers, but rather reflect a system of networked sensors and devices that increasingly monitor everything from earthquakes (Cochran et al., 2009) to cattle (Grobart, 2012) to steps taken by an individual. This chapter examines two interrelated, but distinct, types of digital data generated through sensors: sousveillant and locative. Sousveillant data are data created through self-monitoring via digital technologies (Mann et al., 2003). They are data which are 'consciously employed and controlled by an individual for personal fulfillment', and they have wide overlaps with the Quantified Self movement (Kitchin, 2014). Locative data are derived from any digital media that contain location information; which, in turn, refer to information about places conceived of as points, lines and areas (Tobler, 1979). In modern databases, location information is predominantly stored in terms of x–y point coordinates of latitude and longitude. Any sousveillant device that records location in addition to whatever else it is monitoring – heartrate, steps taken, etc. – is generating locative data. While there is broad overlap between these types of data, they remain distinct; it is possible to have sousveillant data which are not locative, and locative data which are not sousveillant.

Together, a diverse number of scholars have been engaged in and are working on sousveillant and locative media in a variety of sub- and interdisciplinary realms; these include fields as diverse as surveillance studies, information science, critical data studies, medical anthropology, critical geographical information systems (GIS), computational sociology, bio-informatics, urban studies and geo-informatics, among others. To explore the rising significance of these two types of data, the chapter proceeds in three parts. First, it briefly demonstrates the significance of sousveillant and locative data and argues that a key means for understanding their societal significance is through their role as spatial media. Second, it examines current approaches to sousveillant and locative media through the lens of actions and knowledges enabled and constrained (Feenberg, 1999). Finally, it briefly turns to the future, speculatively exploring the forms and roles these spatial media seem most likely to take. This speculation creates a launching point for a necessary and broad discussion on the role these spatial media will play both today and in the future.

SOUSVEILLANT AND LOCATIVE DATA AS SPATIAL MEDIA

From one perspective, the significance of understanding sousveillant and locative data can be argued by volume. Focusing solely on mobile phones, a key generator of spatial data through the use of location-based services (Laurila et al., 2012), and of sousveillant data through the use of pedometers, cameras and other built-in sensors (Reilly, 2013), reveals some staggering figures: for example, consumers purchased approximately 1.3 billion smartphones in 2014 alone (Arthur, 2014). Expanding to mobile devices in general, IBM estimates that two-thirds of the global population keeps one within arms reach at all times, reaching for it 150 times per day (IBM, 2013). In total, this generates 5.2 petabytes of new data a day, creating each week a volume larger than the total amount produced by the Large Hadron Collider in a year (CERN, 2015).

A reliance on sheer volume to justify importance promotes a naïve pseudo-positivistic orientation towards knowledge production, where more is seen to always mean better and to be more meaningful (Wyly, 2013; Kitchin, 2014). While such a position is part of the mythology of 'big data' writ large (boyd and Crawford, 2012), it obscures both the logics behind and functions of sousveillant and locative data within society. As the contributions to this book demonstrate, one means of complicating our understandings of sousveillant and locative data and technologies is to examine how they function in society as spatial media. While it is possible for sousveillant data not to contain geographic information, for the purposes of this chapter, sousveillance is delimited to that which is also locative. Following Crampton (2009), spatial media refer to both new technological objects as well as the spatial data they produce. (in the case of sousveillant and locative media, such objects would include mobile phones, pedometers, heartrate monitors and the like).

Such a move understands media broadly as 'an epistemological approach to digital networked spatial hardware/software objects and information artifacts' that are always necessarily material (Leszczynski, 2015: 746). Couldry and McCarthy (2004: 1) observe that it 'is ever more difficult to tell a story of social space, without also telling a story of media, and vice versa'. As software has come to 'mediate, saturate, and sustain' contemporary capitalism (Graham, 2005: 562), media have similarly come to saturate space (Monteiro, 2014). Spatial media constitute core means through which data are produced, shared and analysed for profit, and are critical components through which society and technology co-mingle in the always-incomplete assemblages through which code saturates and transduces space (Kitchin and Dodge, 2011; Thatcher, 2014).

From the French words for 'below' and 'to watch', sousveillance is a form of 'reflectionism' that, in a specifically socio-political sense, turns surveillance back upon bureaucratic organisations (Mann et al., 2003). In a broader sense, the practice of sousveillance involves the quantification, recording and analysis of an individual's self and environment for means of personal or environmental improvement (Kitchin, 2014; Kitchin and Dodge, 2011). Sousveillant and locative media, then, are one interface that mediates technicity, the 'co-constitutive relations between the human and the technical' (Kinsley, 2013: 13) and, as such, they enable and constrain the actions and knowledges of the individuals who produce, share and analyse them. The remainder of the chapter takes up the double-edged role of spatial media in society, analysing them first in terms of what currently is or has been enabled and constrained, and then what might be so in the future.

CURRENT APPROACHES TO SOUSVEILLANCE

Sousveillant, locative media are interfaces through which individuals communicate, interact and come to know themselves and other individuals, space and their society. Sousveillant and locative media are 'always an effect'; they are 'always a process or a translation' that stand at the boundary between individuals and the world (Galloway, 2012: 33). As such, they function to simultaneously open and foreclose new forms of knowing and politics, new forms of being-in-the-world – that 'irreducible and unsurpassable "embeddedness" [of an individual] in a concrete and ultimately contingent life-world' (Žižek, 1999: 62). Interfaces are attempts to manipulate that being-in-the-world, to control and shape it (de Souza e Silva and Frith, 2012); sousveillant media shape being-in-the-world through two predominant methods – one political and one internal, one based on society and the other on the individual.

As sensors proliferate through society, surveillant objects have become a ubiquitous part of daily life – for example, the inclusion of Radio-Frequency Identification (RFID) tags into common objects makes them potential sites of tracking and surveillance (Gane et al., 2007). Ubiquitous surveillance emerges in a world 'in which it becomes increasingly difficult to escape the proliferating technologies for data collection, storage, and sorting' (Andrejevic, 2012: 92). Mann et al. (2003) posit sousveillance as a direct counter to the increased prevalence of surveillant objects. As originally imagined, sousveillance was inherently social and political in its aims. As a form of 'reflectionism', it was a tactic that followed in the tradition of détournement, of turning the tools and media – or in this case the sensors – of dominant culture back upon itself (de Certeau, 1984; Debord and Wolman, 2006 [1956]).

As interfaces, sousveillant media continue to function in this role today, mediating what can and is known about remote spaces and events. Perhaps the single most provocative and well-known recent examples of sousveillant media mediating public discourse centre around recording police actions in the USA. In places like North Charleston, South Carolina (Schmidt and Apuzzo, 2015), citizens have turned the sensors of everyday life – smartphones – back upon their purported guardians to record and disseminate violent acts. While the focus of this chapter is not upon the specific moral, ethical and legal ramifications of such acts (each issue being extremely worthy of consideration; see Chapters 15 and 22), the acts themselves demonstrate the significant mediatory effects that sousveillant media have in current society. It is also worth noting that this chapter does not address the merits of sousveillant and locative media as forms of citizen science through which the data smart devices produced by transform individuals into passive and active sensors for scientific and urban research (Goodchild, 2007; Haklay, 2013; Resch, 2013; see Chapter 12).

Sousveillant media create spatio-temporal records of events akin to those only previously available through surveillant means. As media, sousveillant media are distributed and redistributed through a variety of channels including social media, broadcast news and online news sources. Created by individuals wielding sensors, once shared, these types of sousveillant media enter into circuits of social and economic capital that shape public discourse, while remaining digital–material–discursive objects co-present at multiple locations and in multiple times. These types of shared sousveillant media transduce spaces through forms of communication and timeless power (Graham et al., 2013) – communication power in how the spatial media produce, create, interpret, recirculate, repackage and contest discourses around themselves; timeless in how sousveillant media continue to exist and (re)present spaces long after the spatio–temporal point they recorded.

The photo of Lieutenant John Pike pepper spraying University of California Davis students on 19 November 2011 serves as an example of the life of such spatial media. The photo presents a sensor–as–interface mediated view of a specific moment in space–time. It was interpreted by social media groups and national media (Seelye, 2011), re-circulated on sites far beyond the original post, repackaged in a variety of mimetic (and memetic) forms including images and t-shirts (Misener, 2011), and contested by socio-political subjects attempting to control discourses around and through it. Four years after the event, the images continue their existence and are still being contested (most recently through attempts by the university to influence their discoverability) – spatial media reminders of a specific point in time that has come to stand in for (to delimit) what is known and said about it. The 'casually pepper spraying cop' helped shape the limits and possibilities of what could be known and said about the socio-technical-political assemblages that were the Davis protests.

Another current form of sousveillant media aims more directly at the self. Often overlapping with what has been termed the Quantified Self movement, this use of spatial media turns the sensors of everyday life inwards to better monitor and, therefore, be able to change aspects of individual life. Considered as spatial media, this type of sousveillant, locative data has a number of interface effects, enabling and constraining what individuals know of themselves and the world around them. This form of spatial media marks a shift in the constitution of surveillance within society: rather than the panoptic metaphor, Boellstorff (2013) suggests the Foucauldian concept of the 'confession' as a 'modern mode of making data, an incitement to discourse we might now term an *incitement to disclose*' (emphasis original). Sousveillant media mediate this confession by quantifying, storing and transmitting data of the self recursively to the individual and outwards to others.

On the one hand, as oft-unseen, corporately controlled algorithms increasingly shape what individuals do and do not see, what they can and cannot know (Dalton and Thatcher, 2014, 2015; Graham et al., 2013), sousveillant media help call into being the 'quantified self-city-nation' (Wilson, 2015). This 'quantified self-city-nation', in its very form, highlights the entangled layers of discourse at which sousveillant spatial media play. For the individual, the constant monitoring of a quantified body provides the insight needed to live a better, more fit, healthier lifestyle – to better enrol within certain discourses of bodily health and well-being. Simultaneously, the data produced through sousveillance are scaled upwards, aggregated, analysed and exchanged as part of the procedures through which new urban and national indicators, dashboards and interfaces are created – by which individuals are then better directed, counted and understood at various spatial scales. These systems less interpret the world than actively frame and produce it in a variety of ways, transducing both spaces and scales of understanding and experience (Kitchin et al., 2015). Algorithmic living – the processes through which 'digital representations and computational algorithms shape our experience of each other' (Mainwaring and Dourish, 2012: 6) – is less a process of interpreting the world than of making it; of making a 'quantified self-city-nation' that emerges as part and parcel of broader processes of neoliberal and capitalist imperatives. Sousveillant media are a chief influence through which algorithms exert influence on everyday life, of both knowing and disciplining individuals and society.

On the other hand, while sousveillant media can and are used by external forces to mediate the interactions between individuals and to transduce in what spaces and at what scales said interactions occur, they are also used recursively by individuals to improve their lives in meaningful ways. For example, a 2013 Pew Research poll (Fox and Duggan, 2013) found that nearly half of those who self-track felt doing so had 'changed their overall approach to maintaining their health or the health of someone for whom they care' (Ramirez, 2013). Danks et al. (2014) found that the use of a StepWatch Activity Monitor significantly improved both the number of steps taken daily and the amount of time spent walking for individuals recovering from a stroke. These real-world benefits are at the heart of the Quantified Self movement, one that believes that the powerful new sensors of the everyday, coupled with powerful algorithms and analysis, can turn the raw data of life into the information necessary for a better life (Wolf, 2010).

QUANTIFIED LIVING NOW AND IN THE FUTURE

While these and other limited examples suggest real and tangible benefits to the recursive use of sousveillant media, the tangled intersections between political, corporate and individual uses of data require explication. The Quantified Self movement began by infusing self-tracking with the 'enthusiasm and optimism of a feel-good guru', focusing on how quantification can and will improve individual lives (Wolcott, 2013). Sousveillance, as originally coined, was meant to use those same sensors for 'surveilling the surveillers' (Mann et al., 2003: 332). The same spatial media may now serve as the interface through which individuals both make themselves and are made known to others, to corporations, and to political and legal bodies: in 2012, an American presidential candidate faced intense backlash and scrutiny after a mobile phone video revealed his derogatory comments about '47 percent' of American citizens (Cillizza, 2013); and a Canadian court in 2014 saw the first introduction of a self-tracking device (a FitBit) as evidence (Crawford, 2014). As the tools of spatial sousveillant media move from independent, self-described 'hacker' culture (Wolf, 2010) to the mainstream mass-marketisation of the quantified self-city-nation (Wilson, 2015), mobile applications such as Waterlogged (reminds individuals when to drink water) and sensors like emWave2 (monitors stress and relaxation) are shifting the enabling, empowering aspects of the quantified self into a constrained 'infantilized self' of algorithmic living (Singer, 2015).

Akin to other technological forms, as the sensors, data and applications of the quantified self attain ubiquity, they retreat from conscious consideration (Dave, 2007). The data once hacked into existence by pioneers like Steve Mann (Shinn, 2001) are now a form of spatial media brought into existence by the sensors of everyday life. Sousveillance becomes surveillant as these data are extracted and analysed as part of massive, multibillion dollar data analytics regimes (Kitchin, 2014; Roderick, 2014), a spatial media-cum-lifestyle tool that re-emerges into conscious consideration only when it fails (Harman, 2010). For example, users of Progressive's Snapshot device – which monitors drivers and offers discounts and penalties based on driving habits – report instances of forgetting the device is installed until damage is done to their vehicle (Horcher, 2014). At the time of writing, these devices remain voluntary; however, there is widespread industry interest in standardising their use. Health insurance agencies see self-monitoring devices as a means of continuously pricing insurance rates, of creating a field of governmentality in which certain actions are immediately rewarded and punished (Tanner, 2013). Before this occurs, in the moments before the flashiness of the quantified self-city-nation seduces society and becomes life's *de rigueur*, there must be a more conscious consideration of sousveillant and locative media. Not just of what they are, but also of what might be – of what is only just coming into being.

One means of doing so is to view spatial sousveillant media as part of a perpetually immanent discourse which mediates the potentialities of lived experience. Thatcher (2013) does so when considering Microsoft's infamous 'avoid ghetto' patent, showing that the same technological form – in this case described in a patent – can invoke vastly different expectations for the future, both hopeful and fear-filled. The spatial media in question, a routing algorithm that would factor in sousveillant media such as previous history and demographic profile, did not need to tangibly exist to inspire a conversation around it: its immanence and possibility were enough. In the same way, the new forms of sousveillant media do not need to exist to engage them critically; waiting for their ubiquity will render such engagements more difficult. Many of the conceptual tools necessary for engagement already exist.

Like the other chapters in this book, this one has focused on sousveillant and locative technologies as forms of spatial media, exploring what they currently enable and constrain for individuals and society. Other approaches to sousveillant media focus on an 'ethics of forgetting' (Dodge and Kitchin, 2007), or more explicitly explore the quantified self-city-nation (Wilson, 2015). Governmentality and biopolitics have also been used productively to think through the quantified self, as has Benjamin's work on abstraction (Sherman, 2015). Studies on surveillance, and particularly emerging economies of surveillance (Crampton et al., 2014), speak to the enrolling of individual bodies into regimes of circulation via sousveillant media (Elwood and Leszczynski, 2011). Each approach reveals different knowledge and focuses on different aspects of the role these new spatial media play in day-to-day life. Each is also necessary to understanding and engaging with sousveillant and locative media in the moments before they disappear.

REFERENCES

Andrejevic, M. (2012) 'Ubiquitous surveillance', in K. Ball, K. Haggerty and D. Yon (eds), *Routledge Handbook of Surveillance Studies*. London: Routledge. pp. 91–8.

Arthur, C. (2014) 'Smartphone explosion in 2014 will see ownership in India pass US', *The Guardian: Global Development*. Available at: http://www.theguardian.com/technology/2014/jan/13/smartphone-explosion-2014-india-us-china-firefoxos-android (accessed 21 July 2016).

Boellstorff, T. (2013) 'Making big data, in theory', *First Monday*, 18 (10). Available at: http://firstmonday.org/ojs/index.php/fm/article/view/4869/3750 (accessed 21 July 2016).

boyd, d. and Crawford, K. (2012) 'Critical questions for big data', *Information, Communication and Society*, 15 (5): 662–79.

CERN (2015) 'Computing: experiments at CERN generate colossal amounts of data. The Data Centre stores it, and sends it around the world for analysis'. Available at: http://home.cern/about/computing (last accessed 21 July 2016).

Cillizza, C. (2013, 4 March) 'Why Mitt Romney's "47 percent" comment was so bad', *The Washington Post*. Available at: http://www.washingtonpost.com/blogs/the-fix/wp/2013/03/04/why-mitt-romneys-47-percent-comment-was-so-bad/ (accessed 21 July 2016).

Cochran, E.S., Lawrence, J.F., Christensen, C. and Jakka, R.S. (2009) 'The Quake-Catcher Network: citizen science expanding seismic horizons', *Sesimological Research Letters*, 80: 26–30.

Couldry, N. and McCarthy, A. (eds) (2004) *MediaSpace: Place, Scale, and Culture in a Media Age*. London: Routledge.

Crampton, J.W. (2009) 'Cartography: maps 2.0', *Progress in Human Geography*, 33: 91–100.

Crampton, J.W., Roberts, S. and Poorthuis, A. (2014) 'The new political economy of geographical intelligence', *Annals of the Association of American Geographers*, 104 (1): 196–214.

Crawford, K. (2014) 'When fitbit is the expert witness', *The Atlantic*. Available at: http://www.theatlantic.com/technology/archive/2014/11/when-fitbit-is-the-expert-witness/382936/ (accessed 21 July 2016).

Dalton, C. and Thatcher, J. (2014) 'What does a critical data studies look like, and why do we care? Seven points for a critical approach to "big data"', *Society and Space Open Site*. Available at: http://societyandspace.com/material/commentaries/craig-dalton-and-jim-thatcher-what-does-a-critical-data-studies-look-like-and-why-do-we-care-seven-points-for-a-critical-approach-to-big-data/ (accessed 21 July 2016).

Dalton, C. and Thatcher, J. (2015) 'Inflated granularity: spatial "big data" and geo-demographics', *SSRN*. Available at: http://papers.ssrn.com/sol3/papers.cfm?abstract_id=2544638 (accessed 21 July 2016).

Danks, K.A., Roos, M.A., McCoy, D. and Reisman, D.S. (2014) 'A step activity monitoring program improves real world walking activity post stroke', *Disability and Rehabilitation*, 36 (26): 2233–6.

Dave, B. (2007) 'Space, sociality and pervasive computing', *Environment and Planning B: Planning and Design*, 34 (3): 381–2.

de Certeau, M. (1984) *The Practice of Everyday Life*. Berkeley, CA: University of California Press.

de Souza e Silva, A. and Frith, J. (2012) *Mobile Interfaces in Public Spaces: Locational Privacy, Control, and Urban Sociability*. New York: Routledge.

Debord, G. and Wolman, G.J. (2006 [1956]) 'A user's guide to détournement'. Available at: http://www.bopsecrets.org/SI/detourn.htm (accessed 21 July 2016).

Dodge, M. and Kitchin, K. (2007) '"Outlines of a world coming into existence": computing and the ethics of forgetting', *Environment and Planning B*, 34 (3): 431–45.

Elwood, S. and Leszczynski, A. (2011) 'Privacy, reconsidered: new representations, data practices, and the geoweb', *Geoforum*, 42 (1): 6–15.

Evans, D. (2011) 'The internet of things'. Available at: http://blogs.cisco.com/diversity/the-internet-of-things-infographic (last accessed 30 May 2016).

Feenberg, A. (1999) *Questioning Technology*. New York: Routledge.

Fox, S. and Duggan, M. (2013) 'Tracking for health'. *PewResearchCenter*. Available at: http://www.pewinternet.org/2013/01/28/tracking-for-health/ (accessed 21 July 2016).

Galloway, A.R. (2012) *The Interface Effect*. Cambridge: Polity Press.

Gane, N., Venn, C. and Hand, M. (2007) 'Ubiquitous surveillance: interview with Katherine Hayles', *Theory, Culture & Society*, 24 (7–8): 349–58.

Goodchild, M. (2007) 'Citizens as sensors: the world of volunteered geography', *GeoJournal*, 69: 211–21.

Graham, M., Zook, M. and Boulton, A. (2013) 'Augmented reality in urban places: contested content and the duplicity of code', *Transactions of the Institute of British Geographers*, 38: 464–79.

Graham, S. (2005) 'Software-sorted geographies', *Progress in Human Geography*, 29 (5): 562–80.

Grobart, S. (2012) 'Big dairy enters the era of big data', *Bloomberg Business*. Available at: http://www.bloomberg.com/bw/articles/2012-10-18/big-dairy-enters-the-era-of-big-data (accessed 21 July 2016).

Haklay, M. (2013) 'Citizen Science and Volunteered Geographic Information – overview and typology of participation', in D.Z. Sui, S. Elwood and M.F. Goodchild (eds), *Crowdsourcing Geographic Knowledge: Volunteered Geographic Information (VGI) in Theory and Practice*. Berlin: Springer. pp. 105–22.

Harman, G. (2010) 'Technology, objects and things in Heidegger', *Cambridge Journal of Economics*, 34: 17–25.

Horcher, G. (2014, 24 November) 'Concerns with insurance devices that monitor for safe-driver discounts', *KIRO 7 News*. Available at: http://www.kirotv.com/news/news/concerns-insurance-devices-monitor-safe-driver-dis/njFb5/ (accessed 21 July 2016).

IBM (2013) 'Mobile isn't a device. It's data', IBM. Available at: https://www-03.ibm.com/innovation/ca/en/smarter-enterprise/perspectives/mobile.html (accessed 2 August 2016).

Kinsley, S. (2013) 'The matter of "virtual" geographies', *Progress in Human Geography*, 38: 364–84.

Kitchin, R. (2014) *The Data Revolution: Big Data, Open Data, Data Infrastructure and their Consequences*. London: Sage.

Kitchin, R. and Dodge, M. (2011) *Code/Space: Software and Everyday Life*. Cambridge, MA: MIT Press.

Kitchin, R., Lauriault, T. and McArdle, G. (2015) 'Knowing and governing cities through urban indicators, city benchmarking, and real-time dashboards', *Regional Studies, Regional Science*, 2 (1): 6–28.

Laurila, J., Gatica-Perez, D., Aad, I., Blom, J., Bornet, O., Do, T., Dousse, O., Eberle, J. and Miettinen, M. (2012) 'The mobile big data challenge', *Nokia Research*. Available at: http://research.nokia.com/files/public/MDC2012_Overview_Laurila GaticaPerezEtAl.pdf (accessed 21 July 2016).

Leszczynski, A. (2015) 'Spatial media/tion', *Progress in Human Geography*, 39 (6): 729–51.

Mainwaring, S. and Dourish, P. (2012) 'Intel Science and Technology Center for Social Computing: White Paper'. Available at: http://www.dourish.com/publications/2012/ISTC-Social-Whitepaper.pdf (accessed 21 July 2016).

Mann, S., Nolan, J. and Wellman, B. (2003) 'Sousveillance: inventing and using wearable computing devices for data collection in surveillance environments', *Surveillance and Society*, 1 (3): 331–55.

Misener, J. (2011, 22 November) 'Etsy artist turns "pepper spray cop" image into peace-advocating t-shirt', *The Huffington Post*. Available at: http://www.huffingtonpost.com/2011/11/22/pepper-spray-cop-shirt_n_1107708.html (accessed 21 July 2016).

Monteiro, S. (2014) 'Editorial: Rethinking media space', *Continuum: Journal of Media and Cultural Studies*, 28 (3): 281–5.

Ramirez, E. (2013) 'Pew internet research: 21% self-track with technology', *Quantified Self*. Available at: http://quantifiedself.com/2013/01/pew-internet-research-the-state-of-self-tracking/ (accessed 21 July 2016).

Reilly, P. (2013) 'The mobile phone: a tool for sousveillance?', *Social Worlds in 100 Objects*. Available at: http://www2.le.ac.uk/projects/social-worlds/all-articles/media-and-communication/mobile-phone (accessed 21 July 2016).

Resch, B. (2013) 'People as sensors and collective sensing – contextual observations complementing geo-sensor network measurements', in J.M. Krips (ed.), *Progress in Location-Based Services*. Berlin: Springer. pp. 391–406.

Roderick, L. (2014) 'Discipline and power in the digital age: the case of the US consumer data broker industry', *Critical Sociology*, 40 (5): 729–46.

Schmidt, M.S. and Apuzzo, M. (2015, 7 April) 'South Carolina officer is charged with murder of Walter Scott', *The New York Times*. Available at: http://www.nytimes.com/2015/04/08/us/south-carolina-officer-is-charged-with-murder-in-black-mans-death.html?_r=0 (accessed 21 July 2016).

Seelye, K.Q. (2011, 22 November) 'Pepper spray's fallout, from crowd control to mocking images', *The New York Times*. Available at: http://www.nytimes.com/2011/11/23/us/pepper-sprays-fallout-from-crowd-control-to-mocking-images.html (accessed 21 July 2016).

Sherman, J. (2015) 'How theory matters: Benjamin, Foucault, and Quantified Self – oh my!', *EPIC*. Available at: https://www.epicpeople.org/how-theory-matters/ (accessed 21 July 2016).

Shinn, E. (2001, 8 July) 'Part man, part machine – all nerd', *The Toronto Star*. Available at: http://wearcam.org/tpw_torontostar/ (accessed 21 July 2016).

Singer, N. (2015, 18 April) 'Technology that prods you to take action, not just collect data', *The New York Times*. Available at: http://www.nytimes.com/2015/04/19/technology/technology-that-prods-you-to-take-action-not-just-collect-data.html (accessed 21 July 2016).

Tanner, A. (2013) 'Data monitoring saves some people money on car insurance, but some will pay more', *Forbes*. Available at: http://www.forbes.com/sites/adamtanner/2013/08/14/data-monitoring-saves-some-people-money-on-car-insurance-but-some-will-pay-more/ (accessed 21 July 2016).

Thatcher, J. (2013) 'Avoiding the ghetto through hope and fear: an analysis of immanent technology using ideal types', *GeoJournal*, 78 (6): 967–80.

Thatcher, J. (2014) 'Living on fumes: digital footprints, data fumes, and the limitations of spatial big data', *International Journal of Communication*, 8: 1765–83.

Tobler, W. (1979) 'A transformational view of cartography', *The American Cartographer*, 6 (2): 101–6.

Wilson, M.W. (2015) 'Flashing lights in the quantified self-city-nation', *Regional Studies, Regional Science*, 2 (1): 39–42.

Wolf, G. (2010, 28 April) 'The data-driven life', *The New York Times Magazine*. Available at: http://www.nytimes.com/2010/05/02/magazine/02self-measurement-t.html?_r=0 (accessed 21 July 2016).

Wolcott, J. (2013) 'Wired up! Ready to go!', *Vanity Fair*. Available at: http://www.vanityfair.com/culture/2013/02/quantified-self-hive-mind-weight-watchers (accessed 21 July 2016).

Wyly, E. (2013) 'The new quantitative revolution', *Dialogues in Human Geography*, 4: 26–38.

Žižek, S. (1999) *The Ticklish Subject*. London: Verso.

6

SOCIAL MEDIA

JESSA LINGEL

Social media platforms offer both social connection and cultural production, linking people to one another and providing the means for identity work, self-promotion and creative play. In a relatively short span of time, these sites have become increasingly integrated into our everyday lives and for many have become an integral part of how we find jobs, shop, interact with our romantic partners, and make sense of current events. As Hine (2014) has argued, the very embeddedness of the internet into everyday life can lead to the obfuscation of the infrastructures that make online platforms possible. In other words, because social media platforms are so interwoven into daily routines, it's easy to overlook the importance of these platforms and how they have in many ways come to shape our expectation of communication and social interaction.

Although in everyday conversation it is common to use the terms 'social media' and 'social network site' (SNS) interchangeably, it is useful to be precise about the different connotations of these terms. 'Social media' is arguably the broadest term, alternately referring to the content posted online, the sites themselves or even the infrastructure supporting online communication. With a more precise (and technically orientated) rubric, boyd and Ellison (2007) defined social network sites as having three distinct characteristics: constructing a public or semi-public profile within a bounded system; articulating a list of other users to whom they are connected; navigating these articulated relations within the site ('Social network sites: A definition', para. 1). By this definition, Facebook, Twitter and LinkedIn are social network sites, while YouTube, Amazon and Craigslist are not, even though these sites are incredibly popular and deeply important sites of social and economic production. As a somewhat broader definition of social media, Light and McGrath (2010) argued that 'SNS may be conceptualized as socio-technical arrangements incorporating technologies that support' social relations and affiliations (p. 291). These sites are also referred to as 'platforms', and Gillespie (2010) has noted the way that this term has political, cultural and technological connotations, which may or may not be readily apparent to users of these sites. boyd (2010) has also described SNSs as 'networked publics', a term that captures both the sense of shared socio-technical space and the practices and norms that emerge through use. Across these definitions, we see an emphasis on people creating representations of themselves (through profiles) and also their social relationships (through networks) via online and mobile displays that are navigable by others.

Within the incredible range of interpersonal functions supported by SNSs – keeping in touch with social networks; telling stories about ourselves, each other and our societies; conducting business needs; promoting creative work; engaging in public discourse and much more – comes a range of opportunities for understanding space. Social media provide explicit descriptions of social spaces and geographic locations (through text, images and video) and also implicit descriptions (through metadata associated with social media, where 'metadata' refers to the data describing the production of data and media, such as the date, time, location and transmission of photos and messages). For a nuanced, holistic understanding of how space is constructed socially and collectively, the practices, traces and documentations of social media platforms are an invaluable source.

ORIGINS

The core elements of social media platforms – existing social networks manifesting through online exchanges, predicated on an ethic of sharing information – have been present from the beginning of internet-based communication, whether through email listservs (see Abbate, 2000) or bulletin board services (see Driscoll, 2014). boyd and Ellison (2007) cluster the development of social media platforms into several waves, beginning with sites that allowed users to articulate their social networks publicly (like SixDegrees.com, Classmates.com and AIM buddy lists). Later, sites like LiveJournal, AsianAvenue, BlackPlanet and MiGente allowed users to create and browse profiles, in addition to developing social connections to other users. Some of these sites are now essentially defunct, while others (like tribe.net and makeoutclub.com) continue to enjoy niche membership, many years after their creation. In the early 2000s, social network sites became increasingly mainstream, with the release of Myspace in 2003 and Facebook in 2004.

Facebook currently dominates the social media playing field, enrolling its first users in 2004. Facebook was originally available only to students at elite universities, before being opened up to college students broadly and then accessible to everyone (or at least, theoretically, everyone over 13). Facebook currently counts over a billion monthly active users, and the majority of this activity (82.8%) occurs outside the USA and Canada (Facebook, 2015). Beyond expanding its user base, Facebook continues to develop tools that extend past the characteristics boyd and Ellison (2007) identified as definitive of SNSs, with functions that range from photo tagging to making user-to-user voice calls. It is partly because of the sophistication of these tools that Facebook is increasingly becoming a stand-in for the internet itself, evidenced by polls that indicate a substantial number of people who claim never to have used the internet yet are dedicated Facebook users (see Levinson, 2015).

And yet, despite this incredible dominance, the playing field for social media popularity shifts constantly. For example, Myspace went from being a dominant social media platform to a punch line, yet in recent years has seen a resurgence. Site popularity is also shaped by regional factors; for example, Weibo is incredibly popular in China (Bamman et al., 2012), and Google-owned Orkut dominated in Brazil (Fragoso, 2006) before its closure in 2014, but these platforms drew few users from outside their respective countries. The geographic boundedness of social media platforms is perhaps surprising, in that one affordance of web-based media is broad accessibility that ostensibly stretches across national borders. Nonetheless, there continue to be stark regional differences in access and literacy that constrain the ability to or benefits of joining one site versus another. For example, Burrell's (2012) ethnography of cybercafé users documented the experiences of young people who were alternately excited by the social connections offered by SNSs and frustrated by the social, cultural and technological hurdles required to participate in online discourses. These disparities in technological access and uptake are simultaneously important for potential research within spatial media and also a necessary factor to consider in any investigation of social media content.

AGENDAS

Social media platforms are largely corporate entities, and the design decisions implemented by the most popular social media sites have important consequences not only for users of that site, but also for those on other platforms, and occasionally the technology industry writ large. From a design standpoint, interventions on one platform may push other sites to respond by developing similar features or policies. Sometimes this ripple effect can benefit smaller competitors; when Facebook came under intense scrutiny over its username policy in the autumn of 2014, the much smaller site Ello enjoyed a surge of popularity as users who sought greater flexibility in identity policy turned away from Facebook's more conservative rules for profiles and naming (Gray, 2014).

In terms of conducting research on geographic information, social media platforms vary in the extent to which they make data accessible to researchers (see Chapters 4 and 5). Most major SNSs offer varying degrees of access to their immense databases of user-generated content via application programming interfaces (APIs), facilitating the development of research projects based on its users. Individual platforms vary in the level of access that they grant to researchers interested in these data, and the relative ease of gathering user data has important consequences for the popularity or feasibility of studying one site over another, meaning that researchers may choose to study one platform over another based at least partly on the ease of accessing data, as opposed to the specific affordances of a given platform in the context of a particular research question. Moreover, it's important to note that, currently, the vast majority of users have their profiles set to default settings that do not report location, calling into question the generalisability or representativeness of these data points. It is possible – but typically very time-consuming – to detect user location even for those who do not report location directly (see Takhteyev et al., 2012), and it is likely that the tools for location detection will continue to improve as spatial media research develops.

Beyond the sites themselves, there are relevant state and federal laws that can shape social media platforms in ways relevant to researchers. In the USA, policy-makers have been reluctant to enact major legislation around social media sites, with the important caveat that law enforcement agencies increasingly incorporate social media data into their investigatory work (see Brayne, 2015); legal action has been far more common in the European Union, as reflected in government advisements on SNS users' data security (Gibbs, 2015) as well as 'right to be forgotten' legislation, which mandates that search engines like Google must respond to requests to have erroneous or prejudicial information removed from query returns (Laursen, 2015). It is also important to note that in circumstances of political instability, access to SNSs can alternatively be used as a tool of dissent (Tufekci and Wilson, 2012; Wolfson, 2014) and of suppression (Rhoads and Fowler, 2011; Whittaker, 2011). Taking a more positive view of government interest in social media, policy-makers have viewed these platforms as an important means of both connecting to constituents (Karpf, 2012; Kreiss and Howard, 2010) and enabling civic participation through initiatives that are often labelled 'e-government' or (in more academic contexts) 'citizen science'. Many of these initiatives involve some component of spatial media, as when residents of a given municipality use geotagging to document development projects (Sturgis, 2015; World Bank Institute, n.d.) or contributing to data-gathering on pollution (Casey, 2014).

LOCATION-AWARE SOCIAL MEDIA

Within the context of spatial media, location-aware mobile technology (LAMT) represents one of the most important developments. Through GPS-enabled smartphones, LAMTs allow users to track their location and activity by 'checking in' at a particular site or venue, and these technologies include the use of apps as well as added functionality of social media sites (Kelley, 2013). For example, Foursquare began as a company that gave users the ability to document their daily lives, access deals from participating vendors and make serendipitous social connections with other users (see Humphreys, 2013). Similar features have since been built into platforms like Facebook and Yelp, and over time, users may gain a sense of status or micro-celebrity through their extended history with a particular location (Schwartz and Halegoua, 2014). As a set of data points, LAMTs offer access to documentation of spaces themselves as well as mobilities between spaces. Scholars have also investigated the unevenness of this type of access across different urban communities (Kelley, 2014; Thatcher, 2013; see Chapter 4).

Related to LAMTs are the various map-based technologies that can be used to create geographic representations of a given dataset, with the most obvious examples being Bing Maps, Google Maps and OpenStreetMaps, as well as the geotagging abilities of sites like Flickr and Twitter. These tools are often gathered under the name 'geoweb', which refers to a set of geospatial technologies and geographic information available on the web (Herring, 1994), where location-based tools, geospatial data and content can be generated and shared by anyone with an internet connection (Roche et al., 2013). Social media researchers may alternately study geoweb productions already in existence (e.g. Stephens, 2013) or adapt their own tools and invite participation as a source of data (e.g. Quercia et al., 2014). In both cases, the data solicited from geoweb projects offer advantages of representing everyday life and experiences for non-expert cartographers and technologists, reflecting the prevalence of mapping technologies across online platforms, and the lowered barrier of entry for engaging with these tools (e.g. Lin, 2013), as well as concerns about surveillance and the changing nature of privacy (Elwood and Leszczynksi, 2011).

Like all technologies, social media platforms are imbued with particular ideologies that may or may not be immediately obvious (Winner, 1989; Wajcman, 2004). Internet scholarship has pointed out the ways in which social media platforms reflect the privilege of their designers (Burrell, 2012; Lingel and Golub, 2015), and critical geographers have long noted the subjective and ideological dimensions of map making (Crampton, 2003; Monmonier, 2010). In order to retain a reflexive stance towards these ideological valences at work in both social and spatial media, analysis can emphasise the multitude of and differences between social media platforms, and also the ethical ramifications of incorporating social media content into spatial media datasets.

Media multiplexity (Haythornwaite, 2005) refers to the ways that different media intervene in our social lives in complex ways. Media multiplexity reminds us that social media platforms do not exist in a vacuum, nor in tidy, static silos. Instead, they are overlapping and mutually constitutive, where the introduction of design features on one site can spill over onto other sites. For example, the now ubiquitous use of the # began as a user-implemented (rather than designer-implemented) feature on Twitter, and has since spread to other platforms. When investigating social media sites, it is important to avoid sectioning off one site from another – indeed, the richest and most nuanced understandings of social media use in everyday life come from a willingness to articulate relationships between sites, practices and communities.

Although social media platforms offer exciting opportunities for spatial media research, these opportunities come with very real ethical concerns, where a key ethical tension in the use of social media data for research is the issue of privacy (see Chapter 22). An incredible amount of highly personal information is published on social media platforms every day; however, the wealth of these data should not be mistaken for a carte blanche invitation to use the data without consent. Even when data are publicly available, it is incumbent on researchers to consider whether users may feel that the use of these data for academic pursuits is ethical. This ethical obligation is necessary partly because many of the current mechanisms for attaining user consent (such as clickwrap user agreements) are inadequate (Hoback et al., 2013). As well there is an issue of context – user data provided willingly in one context (such as the everyday use of social media sites) can become deeply privileged in another context, such as research (Nissenbaum, 2010). Rather than asking whether it is *possible* to undertake a particular social media investigation, it is necessary to ask whether it is *ethical*, whether the various stakeholders involved are informed of how their data will be used and whether they will have the ability to address the representation of those data beyond the context in which they were initially produced. As more and more social functions continue to interweave within and across social media platforms, it will only become more important for researchers to address privacy and other ethical concerns in their investigations of social media.

REFERENCES

Abbate, J. (2000) *Inventing the Internet*. Cambridge, MA: MIT Press.

Bamman, D., O'Connor, B. and Smith, N. (2012) 'Censorship and deletion practices in Chinese social media', *First Monday*, 17 (3).

boyd, d. (2010) 'Social network sites as networked publics: affordances, dynamics, and implications', in Z. Papacharissi (ed.), *Networked Self: Identity, Community, and Culture on Social Network Sites*. London: Routledge. pp. 39–58.

boyd, d. and Ellison, N.B. (2007) 'Social network sites: definition, history, and scholarship', *Journal of Computer-Mediated Communication*, 13: 210–30.

Brayne, S. (2015) 'Stratified surveillance: policing in the age of big data', PhD thesis. Princeton University, NJ. Available at: http://arks.princeton.edu/ark:/88435/dsp012n49t4083 (accessed 2 August 2016).

Burrell, J. (2012) *Invisible Users: Youth in the Internet Cafés of Urban Ghana*. Cambridge, MA: MIT Press.

Casey, M. (2014, 21 November) 'Scientists using social media to track air pollution in China', *CBSNews*. Available at: http://www.cbsnews.com/news/scientists-using-social-media-to-track-air-pollution-in-china/ (accessed 25 July 2016).

Crampton, J.W. (2003) *The Political Mapping of Cyberspace*. Chicago, IL: University of Chicago Press.

Driscoll, K. (2014) 'Hobbyist inter-networking and the popular internet imaginary: forgotten histories of networked personal computing, 1978–1998', doctoral dissertation. University of Southern California, Los Angeles, CA.

Elwood S. and Leszczynski A. (2011) 'Privacy, reconsidered: new representations, data practices, and the geoweb', *Geoforum*, 42: 5–16.

Facebook (2015) 'Facebook company info'. Available at: http://newsroom.fb.com/company-info/ (accessed 25 May 2016).

Fragoso, S. (2006) 'Eu odeio quem odeia ... considerações sobre o comportamento dos usuários brasileiros na "tomada" do Orkut', *Intercom*. Available at: http://compos.org.br/seer/index.php/e-compos/article/viewFile/89/89 (accessed 25 July 2016).

Gibbs, S. (2015, 26 March) 'Leave Facebook if you don't want to be spied on, warns EU', *The Guardian*. Available at: http://www.theguardian.com/technology/2015/mar/26/leave-facebook-snooped-on-warns-eu-safe-harbour-privacy-us (accessed 25 July 2016).

Gillespie, T. (2010) 'The politics of "platforms"', *New Media and Society*, 12 (3): 347–64.

Gray, S. (2014, 26 September) 'Hello Ello! What you need to know about the social network stealing users from Facebook', *Salon*. Available at: http://www.salon.com/2014/09/26/hello_ello_what_you_need_to_know_the_new_social_network_stealing_users_from_facebook/ (accessed 25 July 2016).

Haythornwaite, C. (2005) 'Social networks and internet connectivity effects', *Information, Communication and Society*, 8 (2): 125–47.

Herring, C. (1994) 'An architecture of cyberspace: spatialization of the Internet', *U.S. Army Construction Engineering Research Laboratory*. Available at: http://citeseerx.ist.psu.edu/viewdoc/summary?doi=10.1.1.37.4604 (accessed 25 July 2016).

Hine, C. (2014) *Ethnography for the Internet*. London: Bloomsbury.

Hoback, C., et al. (2013). *Terms and Conditions May Apply*. Sausalito, CA. Distributed by ro★co films educational.

Humphreys, L. (2013) 'Mobile social media: future challenges and opportunities', *Mobile Media and Communication*, 1 (1): 20–5.

Karpf, D. (2012) 'Social science research methods in internet time', *Information, Communication and Society*, 15 (5): 639–61.

Kelley, M.J. (2013) 'The emergent urban imaginaries of geosocial media', *GeoJournal*, 78 (1): 181–203.

Kelley, M.J. (2014) 'Urban experience takes an informational turn: mobile internet usage and the unevenness of geosocial activity', *GeoJournal*, 79 (1): 15–29.

Kreiss, D. and Howard, P.N. (2010) 'New challenges to political privacy: lessons from the first US Presidential race in the Web 2.0 era', *International Journal of Communication*, 4: 1032–50.

Laursen, L. (2015, 23 April) 'How Google handed a year of "Right to be Forgotten" requests', *IEE Spectrum*. Available at: http://spectrum.ieee.org/telecom/internet/how-google-handled-a-year-of-right-to-be-forgotten-requests (accessed 25 July 2016).

Levinson, S. (2015, 9 February) 'Millions of people don't realize Facebook is part of the internet', *Elite Daily*. Available at: http://elitedaily.com/news/technology/millions-people-dont-realize-facebook-part-internet/930519/ (accessed 25 July 2016).

Light, B. and McGrath, K. (2010) 'Ethics and social networking sites: a disclosive analysis of Facebook', *Information Technology and People*, 23 (4): 290–311.

Lin, W. (2013) 'Situating performative neogeography: tracing, mapping, and performing "Everyone's East Lake"', *Environment and Planning A*, 45 (1): 37–54.

Lingel, J. and Golub, A. (2015) 'In face on Facebook: Brooklyn's drag community and sociotechnical practices of online communication', *Journal of Computer-Mediated Communication*, 20 (5): 536–53.

Monmonier, M.S. (2010) *No Dig, No Fly, No Go: How Maps Restrict and Control*. Chicago, IL: University of Chicago Press.

Nissenbaum, H.F. (2010) *Privacy in Context: Technology, Policy, and the Integrity of Social Life*. Stanford, CA: Stanford Law Books.

Quercia, D., O'Hare, N. and Cramer, H. (2014) 'Aesthetic capital: what makes London look beautiful, quiet, and happy?', in *Proceedings of the 17th ACM Conference for Computer Supported Cooperative Work*. New York: ACM Press. pp. 945–55.

Rhoads, C. and Fowler, G.A. (2011, January 29) 'Egypt shuts down internet, cellphone services', *Wall Street Journal*. Available at: http://www.wsj.com/articles/SB10001424052748703956604576110453371369740 (accessed 25 July 2016).

Roche, S., Propeck-Zimmermann, E. and Mericskay, B. (2013) 'GeoWeb and crisis management: issues and perspectives of volunteered geographic information', *GeoJournal*, 78 (1): 21–40.

Schwartz, R. and Halegoua, G. (2014) 'The spatial self: location-based identity performance on social media', *New Media and Society*, 17 (10): 1643–60.

Stephens, M. (2013) 'Gender and the geoweb: divisions in the production of user-generated cartographic information', *GeoJournal*, 78 (6): 981–96.

Sturgis, S. (2015, 19 February) 'Kids in India are sparking urban planning changes by mapping slums', *Citylab*. Available at: http://www.citylab.com/tech/2015/02/kids-are-sparking-urban-planning-changes-by-mapping-their-slums/385636/#site-header (accessed 25 July 2016).

Takhteyev, Y., Gruzd, A. and Wellman, B. (2012) 'Geography of Twitter networks', *Social Networks*, 34 (1): 73–81.

Thatcher, J. (2013) 'Avoiding the ghetto through hope and fear: an analysis of immanent technology using ideal types', *GeoJournal*, 78 (6): 967–80.

Tufekci, Z. and Wilson, C. (2012) 'Social media and the decision to participate in political protest: observations from Tahrir Square', *Journal of Communication*, 62 (2): 363–79.

Wajcman, J. (2004) *TechnoFeminism*. Cambridge: Polity.

Whittaker, Z. (2011, 27 January) 'Egypt "shuts down Internet" amid further protests; Facebook web traffic drops', *Zdnet*. Available at: http://www.zdnet.com/article/egypt-shuts-down-internet-amid-further-protests-facebook-web-traffic-drops/#! (accessed 25 July 2016).

Winner, L. (1989) *The Whale and the Reactor: A Search for Limits in an Age of High Technology*. Chicago, IL: University of Chicago Press.

Wolfson, T. (2014) *Digital Rebellion: The Birth of the Cyber Left*. Champaign, IL: University of Illinois Press.

World Bank Institute (n.d.) 'Promoting citizen feedback through information technologies'. Available at: http://wbi.worldbank.org/wbi/Data/wbi/wbicms/files/drupal-acquia/wbi/citizen_feedback_insert_fy12.pdf (accessed 25 July 2016).

7

URBAN DASHBOARDS

SHANNON MATTERN

We know what rocket science looks like in the movies: a windowless bunker filled with blinking consoles, swivel chairs and shirt-sleeved men in headsets nonchalantly relaying updates from 'Houston' to outer space. Lately, that vision of Mission Control has taken over City Hall. NASA meets Copacabana, proclaimed the *New York Times*, hailing Rio de Janeiro's Operations Center as a 'potentially lucrative experiment that could shape the future of cities around the world'. The *Times* photographed an IBM executive in front of a seemingly endless wall of screens integrating data from 30 city agencies, including transit video, rainfall patterns, crime statistics, data about car accidents and power failures, and more (Singer, 2012; see also Mattern, 2014). These screens constitute spatial media – a spatialised way of seeing, communicating and thinking about the city.

Futuristic control rooms have proliferated in dozens of global cities. Baltimore has its CitiStat Room, where department heads stand at a podium before a wall of screens and account for their units' performance (City of Baltimore, 2002; Pelton, 2002). The mayor's office in London's City Hall features a 4×3 array of iPads mounted in a wooden panel, which seems an almost parodic, Terry Gilliam-esque take on the Brazilian Ops Center. Meanwhile, former British Prime Minister David Cameron commissioned an iPad app – the 'No. 10 Dashboard' (a reference to his residence at 10 Downing Street) – which gave him access to financial, housing, employment and public opinion data. As the *Guardian* reported, 'the prime minister said that he could run government remotely from his smartphone' (O'Hear, 2012; Gibbs, 2014; Newton, 2014).

This is the age of dashboard governance, heralded by gurus like Stephen Few, founder of the 'visual business intelligence' consultancy Perceptual Edge. Few (2004: 3) defines the dashboard as a 'visual display of the most important information needed to achieve one or more objectives; consolidated and arranged on a single screen so the information can be monitored at a glance'. A well-designed dashboard, he says – one that makes proper use of visualisation techniques informed by the 'brain science' of aesthetics and cognition – can afford its users not only a perceptual edge, but a performance edge, too (Few 2004, 2013). The ideal display offers a big-picture view of *what* is happening in real time, along with information on historical trends, so that users can divine the *how* and *why* and redirect future action (see Rasmussen et al., 2009; Nettleton, 2014).

In 2006, when Few published the first edition of his *Information Dashboard Design* manual, folks were just starting to recognise the potential of situated spatial media. Design critic John Thackara (2006: 169) foretold an emerging market for 'global spreadsheets' (his term for data displays) that could monitor the energy use of individual buildings or the ecological footprint of entire cities and regions. Thackara identified a host of dashboard players already on the scene – companies like Juice Software, KnowNow, Rapt, Arzoon, ClosedloopSolutions, SeeBeyond and CrossWorlds – whose names conjured up visions of an omniscient singularity fuelled by data, hubris and Adderall.

By now we know to interpret the branding conceits of tech start-ups with amused scepticism, but those names reflect a recognition that dashboard designers are in the business of translating perception into performance, epistemology into ontology (see Halpern, 2015). They do not merely seek to display information about a system but to generate insights that human analysts use to *change* that system – to render it more efficient or sustainable or profitable, depending upon whatever qualities are valued. The prevalence and accessibility of data are changing the way we see our cities, in ways that we can see more clearly when we examine the history of the urban dashboard.

A HISTORY OF THE DASHBOARD

The term 'dashboard', first used in 1846, originally referred to the board or leather apron on the front of a vehicle that kept horse hooves and wheels from splashing mud into the interior. Only in 1990, according to the *Oxford English Dictionary*, did the term come to denote a 'screen giving a graphical summary of various types of information, typically used to give an overview of (part of) a business organization'. The acknowledged *partiality* of the dashboard's rendering might make us wonder what is bracketed out. Why, all the mud of course! All the dirty (un-'cleaned') data, the variables that have nothing to do with key performance (however it is defined), the parts that do not lend themselves to quantification and visualisation. All the insight that does not accommodate tidy operationalisation and air-tight widgetisation: that is what the dashboard screens out.

Among the very pragmatic reasons that particular forces, resources and variables have historically thwarted widgetisation is that we simply lacked the means to regulate their use and measure them. The history of the dashboard, then, is simultaneously a history of precision measurement, statistics, instrument manufacturing and engineering – electrical, mechanical, and particularly control engineering (see Edgcumbe, 1918; Bennett, 1979, 1993; Mindell, 2002). Consider the dashboard of the Model T Ford. In 1908, the standard package consisted solely of an ammeter, an instrument that measured electrical current, although you could pay extra for a speedometer. You cranked the engine to start it, and once the engine was running, you turned the ignition switch from 'battery' to 'magneto'. There was no fuel gauge until 1909; before then, you dipped a stick in the fuel tank to test your levels.

Water gushing from the radiator was your 'engine temperature warning system'. As new means of measurement emerged, new gauges and displays appeared.

And then things began to evolve in the opposite direction: as more and more mechanical operations were automated, the dashboard evolved to relay their functioning symbolically, rather than indexically. By the mid-1950s, the oil gauge on most models was replaced by a warning, or 'idiot' light. The driver needed only a binary signal: either (1) things are running smoothly; or (2) something's wrong, panic! (Berger, 2001: 240). The 'Maintenance Required' light came to indicate a whole host of black-boxed measurements. The dashboard thus progressively simplified the information relayed to the driver, as much of the hard intellectual and physical labour of driving was now done by the car itself.

Dashboard design in today's automobiles is driven primarily by aesthetics. While some 'high-performance' automobiles are designed to make drivers feel like they're piloting a fighter jet, the dashboard drama is primarily for show. It serves both to market the car *and* to cultivate the identity and agency of the driver: this assemblage of displays requires a new literacy in the language and aesthetics of the interface, which constitutes its own form of symbolic, if not mechanical, mastery.

In an actual fighter jet, of course, all those gauges play a more essential operational role. As Frederick Teichmann wrote, in his 1942 *Airplane Design Manual*, 'All control systems terminate in the cockpit; all operational and navigational instruments are located here; all decisions regarding the flight of the airplane, with … very few exceptions … are determined here' (p. 106). Up through the late 1920s or early 1930s, however, pilots had few instruments to consult. World War I pilots, according to Branden Hookway (2004: 38–9), were 'expected to rely almost solely on unmediated visual data and "natural instinct" for navigation, landing, and target sighting'; navigation depended on a mixture of 'dead reckoning (estimating one's position using log entries, compass, map, etc., in absence of observation) and pilotage (following known landmarks directly observed from the air)'. And while some instruments – altimeter, airspeed indicator, oil pressure and fuel gauges, etc. – had become available by the war's end, they were often inaccurate and illegible, and most pilots continued to fly by instinct and direct sight.

Throughout the 1920s, research funded by the military and by instrument manufacturers like Sperry sought to make 'instrument flying' more viable. By 1928, Teichmann writes, pilots were flying faster, more complicated planes and could no longer 'trust their own senses at high altitudes or in fogs or in cross-country flights or in blind flying'. They came to depend on the dashboard for their survival.

> Before long, the cockpit grew too large for the plane:
>
> Phone lines linked controllers to the various airfields, which communicated with individual planes by high-frequency radio … Plotters hovered around the situation map … A vast electric tableau, glowing in a bewildering array of colored lights and numbers, spanned the wall opposite the viewing cabin like a movie curtain. On this totalizator, or tote board, controllers could see at a glance the pertinent operational details – latest weather, heights of the balloon barrage layer guarding key cities, and most especially, fighter status. (Buderi, 1996: 95)

That was the control room of No. 11 Group of the Royal Air Force (RAF) Fighter Command, at Uxbridge, England, in September 1940, as described by Robert Buderi (1996) in his book on the history of radar. The increasing instrumentation of flight and other military operations, and the adoption of these instrumental control strategies by government and business, led to the creation of immersive environments of mosaic displays, switchboards and dashboards – from Churchill's War Rooms, to the Space Age's mythologised mission control, to the pushbutton controls of the 'Opsroom' in Project Cybersyn, Chile's cybernetics-informed decision-support system designed in the early 1970s to manage the nation's economy (see Medina, 2011).

DASHBOARD DRIVERS

Business and urban data displays often mimic the dashboard instrumentation of cars or aeroplanes. Where in a car you would find indicators for speed, oil and fuel levels, here you will find widgets representing an organisation's 'key performance indicators': cash flow, stocks, inventory and so forth. Bloomberg terminals, which debuted in 1982, allowed finance professionals to customise their multiscreen displays with windows offering real-time and historical data regarding equities, fixed-income securities and derivatives, along with financial news feeds and current events (because social uprisings and natural disasters have economic consequences, too), and messaging windows, where traders could provide context for the data scrolling across their screens. Over the past three decades, the terminals have increased in complexity: they have specialised keyboards and biometric authentication equipment, and 'responsive environments' that are accessible on a variety of devices – not just the iconic two-screen display.

The widespread adoption of the Bloomberg terminal notwithstanding, it took a while for dashboards to catch on in the corporate world. Stephen Few reports that during much of the 1980s and 1990s, large companies focused on amassing data, without carefully considering which indicators were meaningful or how they should be analysed. He argues that the 2001 Enron scandal incited a cultural shift. Recognising the role of data in corporate accountability and ethics, the chief information officers of major companies finally embraced the dashboard's panoptic view.

The dashboard market now extends far beyond the corporate world. In 1994, New York City police commissioner William Bratton adapted former officer Jack Maple's analogue crime maps to create the CompStat model of aggregating and mapping crime statistics. Around the same time, the administrators of Charlotte, North Carolina, borrowed a business idea – Robert Kaplan and David Norton's 'total quality management' strategy known as the 'Balanced Scorecard' – and began tracking performance in five 'focus areas' defined by the City Council: housing and neighbourhood development, community safety, transportation, economic development and the environment.

In 1999, Baltimore mayor Martin O'Malley, confronting a crippling crime rate and high taxes, designed CitiStat, 'an internal process of using metrics to create accountability within his government' (Tauberer, 2014; this rhetoric of data-tested internal 'accountability' is prevalent in early dashboard development efforts). The project turned to face the public in 2003, when Baltimore launched a website of city operational statistics, which inspired DCStat (2005), Maryland's StateStat (2007) and NYCStat (2008) (see Behn, 2007). Since then, myriad other states and metro areas have developed their own dashboards (Kitchin et al., 2015) – driven by a 'new managerialist' approach to urban governance, committed to 'benchmarking' their performance against other regions, and obligated to demonstrate compliance with sustainability agendas.

Some early dashboard projects have already been abandoned, and others have gone on hiatus while they await technical upgrades. The now-dormant LIVE Singapore! project, a collaboration of MIT's Senseable City Lab and the Singapore–MIT Alliance for Research and Technology (SMART), was intended to be an 'open platform' for the collection, combination and distribution of real-time data, and a 'toolbox' that developer communities could use to build their own civic applications. The rise of smartphones and apps has influenced a new wave of projects that seek not just to visualise data but to give us something to do with them, or layer on top of them.

Over the past several years, a group of European cities has been collaborating on the development of urbanAPI, which proposes to help planners engage citizens in making decisions about urban development. Boston's Citizens Connect has more modest aspirations: it allows residents to report potholes, damaged signs and graffiti. Many projects have scaled back their 'built-in' civic engagement aspirations even further. Citizens' agency is limited to accessing data, perhaps customising the dashboard interface and thereby determining which sources are prioritised, and supplying some of those data passively (often unwittingly) via their mobile devices or social media participation. If third parties wish to use the data represented on these platforms in order to develop their own applications, they are free to do so – but the platforms themselves involve few, if any, active participation features.

In 2012, London launched an 'alpha' prototype of the city dashboard that powers the mayor's wall of iPads. Created by the Centre for Advanced Spatial Analysis (CASA) at University College London, and funded by the government through the National e-Infrastructure for Social Simulation, the web-based platform features live information on weather, air quality, train status and surface transit congestion, as well as local news (see Figure 19.1). Data provided by city agencies are supplemented by CASA's own sensors (and, presumably, by London's vast network of CCTV cameras). In aggregate, these sources are meant to convey the 'pulse' of London. Other urban cadences are incorporated via social media trends, including tweets from city media outlets and universities, along with a 'happiness index' based on an 'affect analysis' of London's social media users (O'Brien, 2012). The CASA platform has also been deployed in other UK cities, from Glasgow to Brighton.

By now these dashboard launches are so common that we begin to see patterns. Dublin's dashboard, released in 2014 by the Programmable City project and the All-Island Research Observatory at Maynooth University, integrates data from numerous sources – Dublin City Council, the regional data-sharing initiative Dublinked, the Central Statistics Office, Eurostat and various government departments – and presents them via real-time and historical data visualisations and

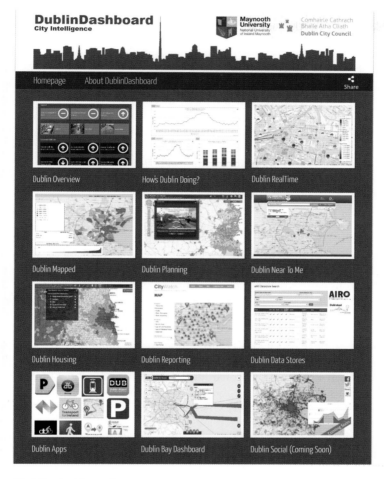

FIGURE 7.1 Dublin dashboard

interactive maps (see Figure 7.1). The platform is intended to help its audiences – citizens, public employees and businesses – with their own decision-making and 'evidence-informed analysis', and to encourage the independent development of visualisations and applications (Kitchin, 2014).

Such projects embody a variety of competing ideologies. They open up data to public consumption and use. They render a city's infrastructures visible and make tangible, or in some way comprehensible, various hard-to-grasp aspects of urban quality of life, including environmental metrics and, in the case of the happiness index, perhaps even mental health. Yet at the same time these platforms often cultivate a top-down, technocratic vision. They also risk perpetuating the fetishisation of data as a 'monetisable' resource and a positivist epistemological unit – and of framing the city as a mere aggregate of variables that can be measured and 'optimised' to produce an efficient or normative system (see Mattern, 2013, 2014).

DASHBOARDING URBAN INTELLIGENCE

Now that dashboards – and the epistemologies and politics they emblematise – have proliferated so widely, across such diverse fields, we need to consider how they frame our vision, what 'mud' they bracket out and how the widgetised screen-image of our cities and regions reflects or refracts the often-dirty reality. We must acknowledge that the dashboard is an epistemological and methodological pastiche. It represents the many ways a governing entity can define what variables are important (and, by extension, what is *not* important) and the various methods of 'operationalising' those variables and gathering data. Of course, whatever is not readily operationalisable or measurable is simply bracketed out. A city's chosen 'key performance indicators', as Kitchin et al. (2015: 7) observe, 'become normalized as a de facto civic epistemology through which a public administration is measured and performance is communicated'.

The dashboard also embodies the many ways of rendering those data represent-able, contextualisable and intelligible to a target audience that likely has only a lim-ited understanding of how the data are derived (see the work of Edward Tufte, and Drucker, 2014). Hookway (2014: 134) notes that 'the history of the interface' – or, in our case, the dashboard – is also a 'history of intelligences … it delimits the boundary condition across which intelligences are brought into a common expression so as to be tested, demonstrated, reconciled, and distributed'. On our urban dashboards we might see a satellite weather map next to a heat map of road traffic, next to a ticker of city expenditures, next to a word-cloud 'mood index' drawing on residents' Twitter and Facebook updates. This juxtaposition represents a tremendous variety of lenses on the city, each with its own operational logic, aesthetic and politics. Viewers can scan across data streams, zoom out to get the big picture, zoom in to capture detail; and this flexibility, as Kitchin et al. (2015: 11) write, improves 'a user's "span of control" over a large repository of voluminous, varied and quickly transitioning data … without the need for specialist analytics skills'. However, while the dashboard's streamlined displays and push-button inputs may lower barriers to entry for users, the dashboard frame also does little to educate those users about where the data come from, or about the politics of information visualisation and knowledge production.

In turn, those representational logics and politics structure the agency and sub-jectivity of the dashboard's users. These tools do not merely define the roles of the user – e.g. passive or active data provider, data monitor, data hacker, app builder, user-of-data-in-citizen-led-urban-redevelopment – they also construct her as an urban subject and define, in part, how she conceives of, relates to, and inhabits her city. Thus, the system also embodies a kind of ontology: it defines what the city *is* and *is not*, by choosing how to represent its parts. If a city is understood as the sum of its compo-nent widgets – weather plus crime statistics plus energy usage plus employment data – residents have an impoverished sense of how they can act as urban subjects.

DASHBOARD GOVERNANCE

For the dashboard's governing users, the system shapes decision-making and pro-motes data-driven approaches to leadership. As noted earlier, dashboards are intended not merely to allow officials to monitor performance and ensure 'accountability', but also to make predictions and projections – and then to change the system in order to render the city more sustainable or profitable or efficient. As Kitchin et al. (2015: 24) note, dashboards seemingly allow for macro, longitudinal views of a city's operations and offer an 'evidence base far superior to anecdote'.

The risk here is that the dashboard's supposed comprehensiveness and seamlessness suggest that we can 'govern by BlackBerry' – or 'fly by instrument' – alone. Such instrumental approaches (given most officials' disinclination to reflect on their own methods) can foster the reification of data and open the door to analytical error and logical fallacy; or, conversely, it promotes the intentional employment of muddy methodology in the pursuit of desirable data (Eterno et al., 2016). As Adam Greenfield (2014) explains:

> Correlation isn't causation, but that's a nicety that may be lost on a mayor or a municipal administration that wants to be seen as vigorously proactive. If fires disproportionately seem to break out in neighborhoods where lots of poor people live, hey, why not simply clear the poor people out and take credit for doing something about fire? After all, the city dashboard you've just invested tens of millions of dollars in made it very clear that neighborhoods that had the one invariably had the other. But maybe there was some underlying, unaddressed factor that generated both fires and the concentration of poverty.

Cities are messy, complex systems, and we cannot understand them without the methodological and epistemological mud. Given that much of what we perceive on our urban dashboards is sanitised, decontextualised and necessarily partial, we have to wonder, too, about the political and ethical implications of this framing: about what ideals of 'openness' and 'accountability' and 'participation' are represented by the sterilised quasi-transparency of the dashboard.

ACKNOWLEDGEMENTS

This chapter is a revised and abridged version of an article that originally appeared as 'Mission Control: A History of the Urban Dashboard', *Places Journal* (March 2015): https://placesjournal.org/article/mission-control-a-history-of-the-urban-dashboard/

REFERENCES

Behn, R.D. (2007) 'What all mayors would like to know about Baltimore's CitiStat Performance Strategy', IBM Center for The Business of Government. Available at: http://web.pdx.edu/~stipakb/download/PerfMeasures/CitiStatPerformanceStrategy.pdf (accessed 25 July 2016).

Bennett, S. (1979) *A History of Control Engineering, 1800–1930*. London: Peter Peregrinus.

Berger, M. (2001) *The Automobile in American History and Culture.* Westport, CT: Greenwood Press.

Buderi, R. (1996) *The Invention that Changed the World: How a Small Group of Radar Pioneers Won the Second World War and Launched a Technological Revolution.* New York: Simon and Schuster.

City of Baltimore (2002, 14 July) 'CitiStat/process/take a tour'. Available at: http://archive.baltimorecity.gov/Government/AgenciesDepartments/CitiStat/Process/TakeATour.aspx (accessed 25 July 2016).

Drucker, J. (2014) *Graphesis: Visual Forms of Knowledge Production.* Cambridge, MA: Harvard University Press/metaLAB.

Edgcumbe, K. (1918) *Industrial Electrical Measuring Instruments,* 2nd edition. New York: D. Van Nostrand Company.

Eterno, J., Verma, A. and Silverman, E.B. (2016) 'Police manipulation of crime reporting: insiders' revelations', *Justice Quarterly,* 33 (5): 811–35.

Few, S. (2004, 20 March) 'Dashboard confusion', *Intelligent Enterprise.* Available at: https://www.perceptualedge.com/articles/ie/dashboard_confusion.pdf (accessed 25 July 2016).

Few, S. (2013) *Information Dashboard Design: Displaying Data for At-a-Glance Monitoring,* 2nd edition. Burlingame, CA: Analytics Press.

Gibbs, S. (2014, 21 August) 'David Cameron: I can manage the country on my BlackBerry', *Guardian.* Available at: http://www.theguardian.com/technology/2014/aug/21/david-cameron-blackberry (accessed 25 July 2016).

Greenfield, A. (2014, 24 February) 'Two recent interviews', *Speedbird.* Available at: https://speedbird.wordpress.com/2014/02/24/two-recent-interviews/ (accessed 25 July 2016).

Halpern, O. (2015) *Beautiful Data: A History of Vision and Reason Since 1945.* Durham, NC: Duke University Press.

Hookway, B. (2004) 'Cockpit', in B. Colomina, A. Brennan and J. Kim (eds), *Cold War Hothouses.* New York: Princeton Architectural Press. pp. 22–54.

Hookway, B. (2014) *Interface.* Cambridge, MA: MIT Press.

Kitchin, R. (2014, 11 September) 'Dublin dashboard launch', *Programmable City.* Available at: http://www.maynoothuniversity.ie/progcity/2014/09/dublin-dashboard-launch-1030-1-00pm-friday-19th-september/ (accessed 25 July 2016).

Kitchin, R., Lauriault, T.P. and McArdle, G. (2015) 'Knowing and governing cities through urban indicators, city benchmarking, and real-time dashboards', *Regional Studies, Regional Science,* 2 (1): 6–28.

Mattern, S. (2013, November) 'Methodolatry and the art of measure', *Places Journal.* Available at: https://placesjournal.org/article/methodolatry-and-the-art-of-measure/ (accessed 25 July 2016).

Mattern, S. (2014, April) 'Interfacing urban intelligence', *Places Journal.* Available at: https://placesjournal.org/article/interfacing-urban-intelligence/ (accessed 25 July 2016).

Medina, E. (2011) *Cybernetic Revolutionaries: Technology and Politics in Allende's Chile.* Cambridge, MA: MIT Press.

Mindell, D.A. (2002) *Between Human and Machine: Feedback, Control and Computing Before Cybernetics.* Baltimore, MD: Johns Hopkins University Press.

Nettleton, D. (2014) *Commercial Data Mining: Processing, Analysis and Modeling for Predictive Analytics Projects.* Waltham, MA: Elsevier.

Newton, A. (2014, 14 March) 'The Number 10 dashboard', *Action 4 Case Study: Digital Capability across Departments*, UK Cabinet Office. Available at: https://www.gov.uk/government/publications/case-study-on-action-4-digital-capability-across-departments/action-4-case-study-digital-capability-across-departments--2 (accessed 25 July 2016).

O'Brien, O. (2012, 23 April) 'City dashboard', *Suprageography*. Available at: http://oobrien.com/2012/04/citydashboard/ (accessed 25 July 2016).

O'Hear, S. (2012, 7 November) 'Well, what do you know: the UK Prime Minister's iPad "app" is real. We have details', *Tech Crunch*. Available at: http://techcrunch.com/2012/11/07/too-many-twits-make-a-pm/ (accessed 25 July 2016).

Pelton, T. (2002) 'Running the city by the numbers', *Baltimore Sun*. Available at: http://articles.baltimoresun.com/2002-07-14/news/0207140246_1_control-room-goggles-mayor (accessed 25 July 2016).

Rasmussen, N.H., Bansal, M. and Chen, C.Y. (2009) *Business Dashboards: A Visual Catalog for Design and Deployment*. Hoboken, NJ: John Wiley and Sons.

Singer, N. (2012, 3 March) 'Mission control, built for cities', *New York Times*. Available at: http://www.nytimes.com/2012/03/04/business/ibm-takes-smarter-cities-concept-to-rio-de-janeiro.html?pagewanted=all&_r=0 (accessed 25 July 2016).

Tauberer, J. (2014) 'History of the movement', *Open Government Data: The Book*, 2nd edition. Available at: https://opengovdata.io/2014/history-the-movement/ (accessed 25 July 2016).

Teichmann, F.K. (1942) *Airplane Design Manual*. New York: Pitman Publishing.

Thackara, J. (2006) *In the Bubble: Designing in a Complex World*. Cambridge, MA: MIT Press.

8

GEODESIGN

STEPHEN ERVIN

Geodesign is a marriage of environmental and landscape planning and design (often including architecture and infrastructure), together with geographic information systems (GIS, see Chapters 2 and 3) and, more broadly, various digital computation and telecommunications technologies, used to inform design decisions and to simulate and evaluate their impacts across a range of criteria. The term is a neologism dating to about 2009, and its promulgation is strongly linked to the GIS software maker Esri, Inc. and their publications (e.g. McElvaney, 2012; Miller, 2012). Geodesign is explicitly a kind of 'computer-aided design', but that term, often abbreviated as 'CAD', has come to have a somewhat-limited meaning: the use of computer software digital tools to make drawings and perhaps 3D models. Geodesign certainly depends on using CAD tools – but its distinguishing characteristic is the use of computing and algorithmic tools (e.g. GIS) to simulate and evaluate impacts, imbuing the process with analysis, not just representation. In practice this means that the subject matter and context for the design needs to justify the additional effort of simulation and impact analysis, and so the usual definition of geodesign includes some notion of a problem large and serious enough to justify its use.

So, geodesign may be simply defined as 'design that is visible on a map', or, as in the subtitle of Carl Steinitz's (2012) recent book on the topic, 'changing geography by design'. This definition, while concise, leaves unmentioned the distinguishing characteristic of impact analysis mentioned above, and seems to focus on the result rather than the process, and so is unsatisfactory.

An earlier definition, attributed to Michael Flaxman, which can be found on Wikipedia, is longer – 'Geodesign is a design and planning method which tightly couples the creation of design proposals with impact simulations informed by geographic contexts.' This is explicitly about impact simulations, to be sure, but still leaves implicit other key aspects of the process. For example, if a problem is large and serious enough to justify computer simulation, then it is likely to be too large and complex for just one designer alone, and will require a multidisciplinary team of designers, which introduces even more complexity and may require computer and communications tools to enable productive collaboration.

Since a new word is only justified when it has real, new and distinct meaning, it seems valuable to have a clear definition that articulates all the important distinctions. To that end, I have previously proposed a somewhat longer and more specific definition:

> Geodesign is environmental design usually involving large areas, complex issues, and multi-person teams, that leverages the powers of digital computing, algorithmic processes, and communications technologies to foster collaborative, information-based design projects, and that depends upon timely feedback about impacts and implications of proposed designs, based on dynamic modeling and simulation, and informed by systems thinking. Systems thinking means that multiple interconnected systems are considered, and that the models and simulations evaluate impacts over a larger area, greater complexity, and longer time-frame than any immediate design proposal. (Ervin, 2012: 158)

This definition requires that digital data, information processing, computing and algorithmic processes are involved (i.e. not just intuition and training); that the efforts are collaborative (not purely individual); that simulations and evaluations are timely (and not only visual); and that the geographic and temporal scope of impacts considered is broad (broader than the 'immediate design proposal'). And 'environmental' design (not 'landscape design') means not just that it happens in the natural and built world, but that the ethics and attitudes of environmentalism are engaged in the process. Finally, requiring 'systems thinking' recognises the dynamic and interconnected nature of the world we live in, and the necessity for identifying and understanding connections and systems (e.g. ecosystems, transportation systems, energy systems and others.) Thus, geodesign projects would ordinarily be aimed at improved environmental health and resilience, not just economic growth or transport efficiency; at sustainability, minimum impact, and concern for ecological structure and function, as well as for human communities and concerns.

Many aspects of the above definition are aspirational – we do not yet have many good examples of full-blown geodesign processes or projects that satisfy all points of this definition. But we already know that all of these aspirations are technologically possible, and the challenge is to bring them to bear on the 'grand challenges' of environmental design in the 21st century, when design and development impacts are literally often global in reach, and problems are immense and complex.

Geodesign efforts and technologies overlap substantially with a field of study called spatial decision support systems (SDSS), aimed at cataloguing, understanding and developing systems to aid in decision-making with a spatial dimension (Wilson, 2015). Clearly, geodesign shares this need. What makes the geodesign project distinct is the explicit inclusion of 'design' – design is not just 'decision-making'; it is also 'option-creating', and engages imagination, creativity and non-linear thinking as much as rational decision-making or problem-solving. This, really, is the key characteristic of geodesign: it is designing and design-thinking, together with geospatial analysis.

ORIGINS

The term 'geodesign' itself arose shortly after a seminal conference held in 2008, at the University of California, Santa Barbara. Sponsored by the National Center for Geographic Information and Analysis (NCGIA), the 'Specialist Meeting on Spatial Concepts in GIS and Design', held 15–16 December 2008, drew together a collection of geographers, designers, computer scientists, GIS technologists and others to deliberate on the question: 'To what extent are the fundamental spatial concepts that lie behind GIS relevant in design?' Or, as the programme suggested, alternatively: 'To what extent can the fundamental spatial concepts of design be addressed with GIS?'

This discussion started with the rather obvious distinctions between 'science' (rational, repeatable, 'left-brain') and 'art' (just the opposite) taken as proxies for 'GIS' and 'design'. Although no consensus was reached, there was considerable interest generated, and nobody suggested that finding a fusion of these undertakings was futile – indeed, participant Jack Dangermond, president and CEO of the aforementioned Esri, Inc. (himself trained in landscape architecture at the Harvard Graduate School of Design), at one point mused: 'If we could find a synthesis of GIS and design, we might call it … geodesign!' Within a year, the term was featured in Esri publications, and the following year, the first annual Geodesign Summit meeting was held in Redlands, California, and they have been held annually ever since. Since 2014, there have also been major geodesign conferences in Asia and Europe, as well (Lee and Dias, 2014). This is not to suggest that 'geodesign' had not been performed before 2009 – but it had certainly not been so labelled.

Some observers have suggested that if geodesign is just environmentally conscious design over a large area, then the regional planner Sir Patrick Geddes at the turn of the 20th century was practising geodesign; and that surely Ian McHarg, whose book *Design with Nature*, published in 1969, articulated a synthesis of geographic and environmental thinking with design-methods ideas, should also qualify. But referring back to the extended definition offered above, neither of these designers used computers, or remote sensing, or telecommunications-enabled collaboration (although McHarg's firm was famous for its multidisciplinary practice and access to specialist collaborators).

In my view the earliest example of geodesign fitting the extended definition may have been in 1967, at the Laboratory for Computer Graphics and Spatial Analysis (LCGSA) at Harvard University, where geospatial computation was just in its infancy, but was already being adopted by Carl Steinitz in his research, teaching and practice. The DELMARVA study of 1967 and the subsequent Honey Hill study of 1969 both exemplified geodesign, in their technical makeup, their collaborative design approach, and their emphasis on impact evaluations and systems thinking (Chrisman, 2006). These academic studies set a precedent for such GIS-enabled design, and Steinitz's career served to educate a cadre of planners and designers, operating worldwide, trained in this approach (although not labelled as 'geodesign' until after his retirement from Harvard).

Fast-forward to 2012, when Esri Press published *Geodesign: Case Studies in Regional and Urban Planning* (McElvaney, 2012). While not all the cases presented met every criterion in the extended definition above, the majority demonstrate the integration of most of the criteria – and clearly demonstrate the idea of GIS-enabled design. Also in 2012, Carl Steinitz published *A Framework for Geodesign: Changing Geography by Design*, which remains the most-cited publication in the field. In this book he expanded upon his previously published 'Framework for Theory Applicable to the Education of Landscape Architects (and Other Environmental Design Professionals)' (1990), in which he laid out six strategic questions which he claimed must be addressed by all environmental design, and which he organised into a six-part framework, starting with 'representation' (data, maps, information about an existing situation), through 'process' and 'evaluation' (relevant natural and other processes and qualities to be considered), to 'change' (plans and designs for a new situation), 'impact' (often measured based on 'processes' identified earlier) and finally 'decision' (in which human, socio-economic and political dimensions come into play.) In the geodesign book, he illustrates the application of this framework and method to a range of case studies and examples, many from his days teaching at Harvard – starting in the mid-1960s, at the LCGSA. In this work, Steinitz identifies four key players, whose combination and interaction are essential to geodesign: 'geographic sciences', 'information technologies', 'design professions' and 'the people of the place'. While not a definition *per se,* this articulation of the essential elements – expanded upon in detail in his book – serves as an excellent illustration of the core concepts of geodesign.

DEVELOPMENT

As mentioned, both the term itself and a number of key publications and conferences have been a joint effort of academia (teachers, scholars, theoreticians) and industry (software creators and vendors, primarily), and to some extent, makers of ancillary equipment, such as surveying, remote sensing, or earth-moving and construction equipment. Some private practitioners – e.g. landscape architects, city planners, building architects, transportation engineers – have been central to the conversation, publications, and theoretical and research-based investigations around geodesign. Others have been quietly practising it, to varying degrees, worldwide. Similarly, municipal and government agencies have been present, though not yet particularly vocal, in the geodesign discussion.

As geodesign builds heavily on the strength of traditional design disciplines – architecture, landscape architecture, civil engineering and urban planning – it is only natural that professional schools teaching these disciplines should take notice. When a new technique or term comes along, each school must evaluate whether and how to incorporate that technique or term into its curriculum. As geodesign is both term and technique and is relatively young, the adoption by schools is mixed. Some may take the stand 'We already teach geodesign, and we don't need a new term for what we do'. Others may simply see a new development in practice they need to cover in their curriculum, while yet others may see a competitive advantage in adopting a differentiating term or technique. In the USA, there are already a number of state and private universities offering degrees – undergraduate, graduate, online and traditional – and certificates in geodesign (Penn State, Northern Arizona University,

University of Southern California, Philadelphia University, Arizona State University, among others). Their agendas may range from saving the world, to covering the professions objectively, to blatant public relations and marketing. It is sometimes hard to tell which agenda motivates these programmes, and often they seem to be a mixture of all three.

Software vendors are necessarily more commercial in their outlook. More geodesign means more computers and software in use, and therefore more sales. Especially if their software is uniquely well-positioned for use in geodesign projects, they then have a motivation to spread the word about geodesign. They may also have other motivations, too, of course – just as with schools, their motivations may be mixed. (It is not clear that sales departments are any more likely to delight in an invented word than are academic professors.)

Practitioners similarly need to combine idealism with business needs. Having geodesign expertise and services, and language, to sell in a world that desperately needs them, is a reasonable business proposition. Practitioners are the most likely to be empirical and prepared to compromise on details of definition, but are also the most likely to make real contributions and incremental improvements to the practice of the discipline.

Governmental, municipal and non-governmental groups and agencies are also key players. As geodesign projects often engage large areas and complex projects, there are likely to be regulatory, conservation, infrastructural, legal and other concerns. At present, the term 'geodesign' has no particular legal status or implication (whereas 'architect', 'landscape architect', 'civil engineer', etc. are covered by regulations and registrations, varying somewhat from jurisdiction to jurisdiction). To the extent that professional practitioners and government contractors adopt the tools and terminology of geodesign, it is possible that some aspects of the discipline may become more codified in laws or regulations in the future. All of the above actors are evident in the geodesign community, as represented by the annual Geodesign Summit conferences held to date, in the USA, Asia and Europe (Lee and Dias, 2014).

GEODESIGN AND SPATIAL MEDIA

While geodesign is not a spatial medium in and of itself, it is nonetheless a practice that depends upon spatial media. Spatial media – understood as the whole panoply of digitally enabled devices, processes, networks and communities, and their underlying 'location-aware' and geospatial technologies – are essential to geodesign. Geodesigners must exist in a complex system of systems: natural systems; network and software and telecommunications systems; social, political and economic systems, and others. And geodesign teams must be critical and effective consumers (and producers) of spatial media. Geodata must not only be gathered, but critically evaluated and carefully transformed, through both technical and social means, into information. Representations must be absorbed, transformed, produced and critically examined. It is usually assumed that geodesign projects are undertaken with the best of intentions, and guided by clear environmental and social ethics, but because of the complexity of geodesign problems and the underlying spatial media, it is easy to imagine ways the outcome of a geodesign process could be biased or undesirable, especially to one group or another. Selectively ignoring or distorting data, limiting the range and number of alternatives considered or proposed, and forcing a preconceived solution or denying logical consequences or predicted impacts are all possible with geodesign (although not necessarily any more so than with another kind of design – and they are in any event surely to be labelled 'bad geodesign').

Like all design projects, geodesign projects must confront numerous seemingly irreconcilable challenges. Crowd-sourced data may need to be integrated with industrially produced datasets; multiple stakeholders in a community will have to be heard, and addressed; and quantitative and rational decision criteria may be overwhelmed by political or emotional forces. The requirements of 'design thinking' – making sense of multiplicity, creating collaboratively, embracing diversity, choosing from infinite options, to name just a few – are daunting, and made more so by the rapidly changing nature of spatial media: new sensors, new formats, new protocols, new vendors, new science, new devices and new mashups, all appearing at seemingly exponential rates.

As with any new media, the possibility of an uneven fit with existing media, or processes, or even worldviews, is strong. To evaluate these dependencies and vulnerabilities in geodesign, it may be helpful to step through the ten requirements of my extended definition above and consider the kinds of media literacy that may be required, and interface effects that may be encountered.

'GEODESIGN IS ENVIRONMENTAL DESIGN ...'

1. '... *usually involving large areas* ...' Large areas may be represented by a single synaptic view (e.g. satellite image), or an overlapping mosaic (many individual visitors' photographs). How are large area data collected, and what power structures do they reveal and repress? How can we evaluate their accuracy and reliability? What are the advantages and disadvantages of different types of representations? What outcomes do they potentially predispose, or effectively discriminate against?

2. '... *complex issues* ...' Complex issues by definition have multiple sides, viewpoints and stakeholders. How are these chosen and represented? How is complexity reduced by abstraction and simplification, and what effects do these have on the results?

3. '... *and multi-person teams* ...' Multi-person teams bring both extra brain-power and potentially broader conceptions, but also the possibilities of personality conflicts and power imbalances. How will these be detected? Guarded against? And accounted for? How can social networks assist?

4. '... *that leverages the powers of digital computing* ...' Digital computing offers many powers, such as combinatorial repetition, fast search, fuzzy logic and complex chains of reasoning, but what tools and training are required to make use of them well? Also, what are the notable deficiencies in digital computing (e.g. the ability to reason metaphorically, or in abstractions), and how can they be overcome?

5. '... *algorithmic processes* ...' Algorithmic processes offer modularity and the possibility of parametric variation, but they are vulnerable to hacking and

insidious logic bugs, and may depend upon proprietary syntax and semantics. How can flaws be detected or prevented? How can code be shared, and 'mash-ups' made most valuable?

6. '… *and communications technologies* …' From one-on-one video-conferencing to anonymous broadcast tweets, communications technologies offer a plethora of options. How are these to be evaluated? Deployed? Best used?

7. '… *to foster collaborative, information-based design projects* …' Information-based design may be a two-edged sword: if the information is flawed, so may the design be ('garbage in, garbage out'). How can the quality of information and of collaboration be monitored and improved? What is the role of real-time information in geodesign?

8. '… *that depends upon timely feedback about impacts and implications of proposed designs* …' Impacts and implications may be experienced quite differentially by different individuals, communities and agents. How can impact evaluations be tailored to the realities of the community? What is the role of 'experts' in this? How can 'virtual-' and 'augmented reality' be effective in conveying geodesign?

9. '… *based on dynamic modeling and simulation* …' Dynamic simulations can be difficult to perform and evaluate, and may yield only approximations. How can they be calibrated and understood? How are they susceptible to being hacked, and how can they best be represented, mashed up and repurposed?

10. '… *and informed by systems thinking.*' Systems thinking is a broad and complex challenge, and difficult to quantify and identify. How can it be encouraged and enabled? What digital tools, techniques and media can be used to reveal and amplify systems thinking?

One of the biggest concerns with all 'computer-aided design' is that it may be robotic, inhuman, unfeeling. The core question addressed by the University of California Santa Barbara Spatial Concepts symposium in 2008 is still incompletely answered: 'What are the possibilities for synergy between GIS and design, and where are there incompatibilities?' These possibilities will largely be revealed empirically, by practice, rather than by belief or supposition. As geodesign practice becomes more widespread, it will be more likely that excellent and inspiring examples will appear, which integrate spatial media, modern engineering and social concerns. A community of people using technology and spatial media is a very different prospect to a satellite sensor or a robotic designer working alone; the question is, how can each best be used? To which the answers will likely be many, and varied, and surely regionally and culturally differentiated.

Will a spatial-media-savvy collaborative geodesign team be less susceptible to the 'garbage in, garbage out' concern of basic computer processing? Will they be more or less likely to resort to 'lying with maps'? Will they be able to develop effective internal detection and correction abilities, and more holistic communication capacities? These questions need to be revisited regularly as the geodesign project evolves.

CONCLUDING THOUGHTS

Advanced issues in the practice of geodesign can be roughly divided into two categories: technology-orientated (GIScience) issues and the human-orientated (design) issues. Advanced technology issues with respect to geodesign have broadly to do with technologies for 'sensing' (e.g. satellites, cellphones), for 'simulating' (including visual and other kinds of simulations), for 'reasoning' (such as design software tools, or artificial intelligence) and for 'effecting' (such as lights, motors and all kinds of fabrication technologies). As all of these groups of technologies inevitably develop greater sophistication, power and reach, their applications in geodesign will continually be evolving. Can we design not just static structures based on remote sensing, but dynamic structures that respond to local sensors (e.g. to react to floods, winds, and other environmental changes and hazards?) Can we embed sensors, effectors and reasoning capacity into building materials, for more effective, more responsive, environments?

On the human/design side, there remains a great lack of understanding of how best to engage communities and individuals in collaborative design processes. How can designers inform communities of issues and options? And vice versa? What kinds of simulations and impact analyses are most effective, under what circumstances, to what kinds of groups of people? As the ubiquitous technological 'ether' in which we live and breathe evolves (smartphones, smartcars, smarthouses, smartcities, etc.), how do we best integrate and engage human intelligences and feelings with design tools to achieve real 'computer-aided design'? That is the challenge, and promise, of geodesign as one form of spatial media.

REFERENCES

Chrisman, N. (2006) *Charting the Unknown: How Computer Mapping at Harvard became GIS*. Redlands, CA: Esri Press.

Ervin, S. (2012) 'A system for geodesign', in E. Buhmann, M. Pietsch and S. Ervin (eds), *Proceedings of Digital Landscape Architecture 2011 and 2012 of Anhalt University of Applied Sciences*. Berlin: Wichmann.

Lee, D. and Dias, E. (eds) (2014) *Geodesign by Integrating Design and Geospatial Sciences*. Dordrecht: Springer.

McElvaney, S. (2012) *Geodesign: Case Studies in Regional and Urban Planning*. Redlands, CA: Esri Press.

Miller, W. (2012) *Introducing Geodesign: The Concept*. Redlands, CA: Esri Press.

Steinitz, C. (1990) 'Framework for theory applicable to the education of landscape architects (and other environmental design professionals)', *Landscape Journal*, 9 (2): 136–43.

Steinitz, C. (2012) *A Framework for Geodesign: Changing Geography by Design*. Redlands, CA: Esri Press.

Wilson, M.W. (2015) 'On the criticality of mapping practices: geodesign as critical GIS?', *Landscape and Urban Planning*, 142: 226–34.

PART 2:
SPATIAL DATA
AND SPATIAL MEDIA

9
OPEN SPATIAL DATA

TRACEY P. LAURIAULT

INTRODUCTION

Without spatial data there are no spatial media. Spatial data are understood as any data (quantitative and qualitative) which have a location (e.g. spatially referenced with coordinates) or topology. Spatial data can concern any phenomena and be stored in a variety of formats (e.g. vector, raster, text, video); structured or not; big or small in nature; disseminated via any platform, software or application; distributed under a variety of licence regimes (i.e. open or closed); and produced by individuals, communities (e.g. indigenous, OpenStreetMap), corporations, non-governmental organisations (NGOs) and government institutions alike. They can be exhaust data or metadata from a device or sensor, or generated by an algorithm, and collected, used and reused by experts and amateurs, professionals or scientists, under any type of ideological regime or according to any particular cultural predisposition.

Spatial data can be captured and/or produced by a diverse set of technologies. These include those generated by classic geospatial techniques and technologies (e.g. surveyors, radar, geographic information systems (GIS)), but also wearables, mobile communication devices, in-car navigation systems, traffic cams, satellites, drones, sensors, the internet of things (IoT), innumerable geomatics technologies, and location-enabled social media platforms and applications. Also included are big data tracking and scanning devices in retail, production and shipping, or loyalty cards, credit cards and smart meters, and all the spatial data generated from surveillance systems such as biometric screening and access controls (see Chapter 20).

Captured and volunteered spatial data may be extracted, described, annotated, modelled and stored in myriad databases; fused, linked, mapped and transformed into indicators (see Chapter 11); visualised in dashboards and control rooms (see Chapter 7); and then re-disseminated in webmapping services, geodemographic tools, via application programming interfaces (APIs), in portals, and trusted digital repositories or as linked data. In some cases, real-world objects are topologically stored in geospatial databases and no longer need to be fixed into map layers, as all elements that constitute a map get fused together on the fly with other datasets via geographic unique identification codes (GUIDs). Ordnance Survey Ireland (OSi), for example, now structures its spatial data in such a way that they can be joined with the land registry, land valuation, postcodes, utilities and inventory management systems to manage spatial intelligence and flow (e.g. its Prime2 database). This database also feeds into its spatial media offerings such as GeoHive. They display some of the characteristics of big data with real-time updating, indexicality and relationality and, when linked with other datasets, volume (Kitchin and McArdle, 2016; McNerney, 2016).

Spatial data are also embedded into autonomous systems, machine-learning environments and decision support systems, and can be directed to and/or autonomously drive decision-making and innumerable actions (e.g. robots, marine navigation). Light detection and ranging (LiDAR) point clouds, for example, are now being generated along roads to create 3D meshed environments to enable autonomous cars (Riofrio, 2016), and combined with imagery to develop z-elevation coordinates for 3D simulations (Holland, 2015). Exhaust spatial data and metadata on the other hand can be scrubbed down to their bare spatial attributes to reveal their location and pattern, as Uber users (Mueffelmann, 2015), porn viewers (Robbins, 2015) and Ashley Madison subscribers (Robinson, 2015) have discovered.

Spatial data are also not new. Spatial data have been collected for centuries: we need simply think of the Domesday Book of 1086,[1] the maps painstakingly drawn by explorers to navigate to the 'new' world, and the data they brought back about the places they 'discovered'. In some cases spatial data preceded and made nations (Winichakul, 1994), while in other instances empires were forged in the minds of colonised subjects with a thick red line around claimed 'empty' territories,[2] followed by the fuzzy lines of land treaties[3] and the precise geometry of the surveyors' grids.[4] Colonised territories and new nations were then governmentally and biopolitically managed (Foucault, 2007), with data and maps playing a key role in this ongoing process (Lauriault, 2012). The shift from analogue to digitally mapped spatial data starts in the 1960s with the Canadian Geographic Information System (CGIS) (Fisher and Macdonald, 1979) and work at the Harvard Laboratory for Computer Graphics and Spatial Analysis (Chrisman, 2006). Subsequently, 'old'-media spatial data are often digitised with paper maps scanned, while many historical logs, such as climate data and surveyor notations, are typed into databases. Often these historical records are re-spatialised into spatial media platforms, such as cybercartographic atlases (online, interactive mapping platforms) (Taylor and Lauriault, 2014), while oral cultures and local and traditional knowledge are remediated to become spatialised records (see Chapter 13).

Government geomatics institutions and scientists are generally regarded as being the first to share their data openly, making them accessible: as part of spatial data infrastructures (NRC, 1993; Coleman and McLaughlin, 1997); within national atlases and mapping initiatives, with the first online national atlas released in 1999 (NRCan, 2015); and in specialised data portals (Minnesota Governor's Council on Geographic Information, 1997; Wilson and O'Neil, 1999), webmapping applications (Scharl and Tochtermann, 2007), live webcams, or as part of mapping platforms to visualise and discover open administration data (e.g. Geohive.ie). Spatial framework data, for example, typify this openness. These were considered (in the mid- to late-1990s) as a sensible way to ensure that geospatial data users 'sing from the same song sheet', since using the same authoritative political boundary, road network, hydrographic, and geodetic reference data ensures comparability and interoperability and adheres to the 'create once and use many times' philosophy of authoritative mapping institutions. This also typifies the non-rivalrous nature of digital data.

It is therefore safe to say that there is no shortage of spatial data. However, not all spatial data are equal in terms of quality, usability, accessibility and openness, nor do all systems that employ spatial data constitute spatial media. That said, all spatial media are reliant on spatial data and associated infrastructure. It is also true that the release of certain spatial datasets might be considered as the primary enablers of spatial media. The release of global positioning systems (GPS) for civilian use, for example, has been a key driver of spatial media, along with many government framework data. From GPS comes synchronised time, location and mobility, and with framework spatial data come standardised geometry and features. It is essential for GPS and framework spatial data to be used in spatial media, and for spatial media data to be repurposed for other uses (such as locative media data being used in dashboards), and that they are open or at least accessible through APIs.

OPEN SPATIAL DATA

Data can be framed technically, ethically, in a political economy, spatially and temporally, and philosophically (Kitchin, 2014). In the previous section, spatial data were mostly framed technically, and to some extent historically and philosophically. In this section, spatial data are prefixed with the term 'open', leading to a more nuanced ideological, philosophical and ethical discussion. This section situates spatial data within the discursive regimes of open data and access to data, placing them genealogically along a continuum of a select number of 'open' definitions. Throughout the following discussion it is important to keep in mind that access to data is not synonymous with open data; data can be made available to certain groups for certain usage, rather than to everyone on an unrestrictive basis.

OPENNESS

Open data is a discursive regime which promotes the absence of restrictions on access to publicly funded administrative, government and scientific data[5] in terms of cost, licences, formats, the types of use, whether or not the data are machine readable and in some instances if the data are linked (see Figure 9.1) (OKF, 2005; Berners-Lee, 2006; G8, 2013). While the antecedents of open data have a long history, the rapid rise of the open data movement coincides with the development of the spatial web, most notably with the launch of Google Maps and resulting mashups, Web 2.0 platforms such as Facebook and Wikipedia, and related spatial data and media cultures such as neogeography (e.g. OpenStreetMap), and the UK *Guardian*'s 'Give us back

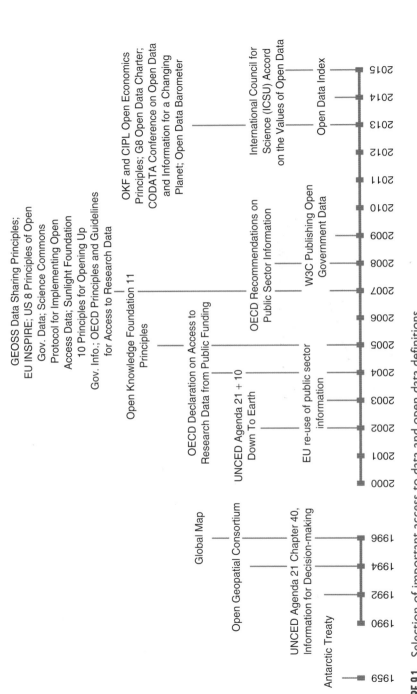

FIGURE 9.1 Selection of important access to data and open data definitions

our crown jewels' campaign to open the UK Ordnance Survey spatial data store (Arthur and Cross, 2006).

The Open Knowledge Foundation is one of the prominent international open-data civil society organisations; based in the UK, it has numerous chapters globally. Its *11 Principles of Open Data* (2005) are one of the earliest statements concerning open data, whereby data are considered open when 'data and content can be freely used, modified, and shared by anyone for any purpose' (2016). Open data as a socio-technological movement has been successful in advocating for the release of data, advancing issues surrounding data policy and management, standards for some thematic datasets (e.g. International Aid Transparency Initiative (IATI)) and, to a lesser extent, data politics.[6] It has made data more visible to citizens and civil society organisations too, while also promoting data-informed public engagement, although with uneven results. Most open data definitions are reductive and are technological framings of data, focusing primarily on the qualities of data, but not necessarily data quality when compared to the access to spatial data discourse. The discourse supporting the opening of data is also generally situated around innovation, the economic return of application development and the reduction in cost resulting from greater government efficiency (MGI, 2013). Transparency and evidence-based decision-making as ideals are present, but they are not always at the forefront, while the private sector is developing a growing interest in capitalising on 'free' government data in a number of ways, such as benefiting from corporate tax incentives (Dalby and Scott, 2015) and being able to freely capitalise on a state-subsidised resource (Gartner, 2015; ODX, 2016). This includes multinational organisations who mine national datasets from nations they have not made any contributions to, with the aim of extracting value from them. A cursory examination of national open data portals reveals that the bulk of the data within these are spatial data, since spatial data were generally already accessible on the internet prior to the emergence of open data discourses, and were already standardised and disseminated in technologically robust interoperable systems and infrastructures.

National and subnational open data initiatives are also compared and ranked against each other in international indicator systems, with the most popular being the Open Data Index (OKF, 2014) by the Open Knowledge Foundation, the Tim Berners-Lee 5 Stars Ranking of Open Data (2006) and the Open Data Barometer (2016), the latter of which is part of an international development research project. In all three cases, initiatives are ranked according to how 'open' data are in technical terms and, except for the Barometer, there is very little discussion about the underlying contextual elements, such as the quality of the internet infrastructure, numeracy or the political economy within which the data are situated. For example, why are Taiwan, the UK and Denmark the top three in the Index, and Libya, Syria and Myanmar at the bottom? Furthermore, critical debate about the robustness of the Index as a system and its methodology (including the comparability of results across time when definitions change, and the validity of ranking when data types and countries are added) is invisible to the public even though debated among volunteers and experts on listserves. The 5 Star Ranking on the other hand, is much more reductive, as it focuses on data formats, with the top 5 Star Ranking being linked data, a technological process that is still in an R&D phase and is relatively inaccessible to most.

Recently, the private sector has become an actor in open data evaluation, with examples such as Open Data 500, which is a 'comprehensive study of U.S. companies that use open government data to generate new business and develop new products and services' (2016), and the Open Cities Index, which sell their benchmarking tool to Canadian municipalities (PSD, 2015). Ironically, the methodologies and algorithms of these two private-sector systems remain closed, and in this case, data about open data are not open as these are behind paywalls, under exclusive copyright and considered to be corporate IP.

Open data is also associated with open government, and nations that are members of the Open Government Partnership sign a declaration committing to 'increase the availability of information about governmental activities' (OGP, 2011) and include open data as part of their national plans (see Chapter 18). A number of countries have also signed on to the International Open Data Charter (2016), agreeing to make their data 'open by default'. Open data is now a key aspect of national and subnational digital strategies; there are open data directives, policies, plans, road-maps and guidebooks. Governments generally, however, have not been critically reflexive of the quality of their programmes and their practices, especially if they rank highly on indexes and meet international or local benchmarks. Canada, for example, in the past few years generally ranked quite highly on both open data and open government indicator systems, yet at the same time it was cancelling the long-form census, muzzling scientists, closing environmental-monitoring stations, shutting science libraries and creating a chilly climate for civil society organisations that did not agree with government policy on race, gender and the environment. It is therefore possible to have undemocratic tendencies and have a great open data record. Alternatively, civil society advocates are organising national summits in collaboration with the private sector and government, at times walking a fine line between being critical and being co-opted. Concurrently, a number of social open data enterprises have emerged (e.g. Ajah, 2016).

Overall, open data as a movement has done much good, but it has not yet met its ideal state where evidence-informed and participatory policy-making is the norm, and politically critical data or data about social and health issues are accessible, although the International Charter and the Barometer are nudging nations in the right direction (but leaning more towards technocratic, not deliberative, solutions). Indeed, it is important to remember that open data has a discursive regime, expresses a form of power/knowledge, performs governmentality and it is normalising as an institution with its own somewhat circular knowledge-producing processes. Moreover, open data, despite appearances and absences in the narrative of its practitioners, did not suddenly appear out of nowhere in 2005; it has a past, and it is associated with established authoritative institutions striving to make data – especially spatial data – accessible on a large scale, for specific purposes.

ACCESS

The regime that promotes 'access to data' includes international natural and social science researchers, environmentalists, earth observation and geomatics communities, governments with spatial data infrastructures, and transnational organisations, such as the Organisation for Economic Co-operation and Development (OECD) and the United Nations (UN) engaged in international and sustainable development, or the European Union (EU) engaged with the facilitation of the regional integration of national data assets across borders (see Figure 9.1). It differs from 'open data' discursively, as it is more about data sharing between certain actors, and has a different data culture and related subcultures. In particular, the narrative on access to scientific and spatial data is more systemic, institutional, organisational, collaborative, research- and results-based, and targeted for a particular purpose. Moreover, the data can be held in proprietary formats and be accessed using a licence. Like open data, the access-to-data discourse has technocratic leanings; although there were no democratic or citizen engagement promises, it is nonetheless about state and scientific institutions collaborating to share and standardise data towards common goals and for specific purposes and not simply for the sake of openness. Principles are about sharing data for sustainable development, resource management, evidence-based decision-making in those areas and the 'benefit of mankind' in somewhat grandiose terms. On a surface level this discursive regime seems more attuned to the data cultures of geomatics, GIS and spatial data infrastructure (SDI); however, a deeper look points to it being the underlying data and infrastructure for spatial media as discussed in Chapter 1, especially if the Open Geospatial Consortium is thought of as the geo equivalent to the World Wide Web Consortium (W3C, 2016), which is responsible for web and spatial media standards.

The Antarctic Treaty of 1959 is a useful and illustrative starting point to demonstrate these differences as seen in Article III below:

1. In order to promote international cooperation in scientific investigation in Antarctica, as provided for in Article II of the present treaty, the Contracting Parties agree that, to the greatest extent feasible and practicable:

    (a) information regarding plans for scientific programs in Antarctica shall be exchanged to permit maximum economy and efficiency of operations;

    (b) scientific personnel shall be exchanged in Antarctica between expeditions and stations;

    (c) scientific observations and results from Antarctica shall be exchanged and made freely available.

As a result of this treaty the Scientific Committee on Antarctic Research (SCAR) 'promotes free and unrestricted access to Antarctic data and information by promoting open and accessible archiving practices' (2016), while the research programmes in some countries such as Canada associated with the International Polar Year (IPY Canada, 2010) stipulated as part of the granting process that data were to be managed and deposited in an archive to ensure open access and future reuse.

Another key milestone in making spatial data accessible (though not necessarily open) is the United Nations Earth Summit of 1992. Chapter 40 – 'Information for decision-making' – mandated nations to collect and manage their geospatial data and information assets, but also to openly share them, as well as build the capacity of all nations to use them. Chapter 40 (UN, 1992) opens with the following statement:

> 40.1. In sustainable development, everyone is a user and provider of information considered in the broad sense. That includes data, information, appropriately packaged experience and knowledge. The need for information arises at all levels, from that of senior decision makers at the national and international levels to the grass-roots and individual levels. The following two programme areas need to be implemented to ensure that decisions are based increasingly on sound information:
>
> a. Bridging the data gap;
>
> b. Improving information availability.

It is important to recognise that this is the precursor to what we now recognise as crowdsourcing, Volunteered Geographic Information (VGI), prosumption and the normalisation of local and traditional knowledge as a form of data. UN Agenda 21+10, *Down to Earth: Geographical Information for Sustainable Development* held in South Africa (NAS, 2002), extends the notion of access to more than data and information and includes access to: education, skills, software, standards, spatial data infrastructures, telecommunication infrastructure, and earth observation systems and related imagery. The Global Earth Observation System of Systems (GEOSS) Principles (2007) build upon these two conferences while implementing and operationalising its geodata sharing policies. This is a significant step forward when considering that Earth observation data underpins a multibillion dollar private- and public-sector global technology partnership and, in this case, one 'that envisions "a future wherein decisions and actions for the benefit of humankind are informed by coordinated, comprehensive and sustained Earth observations and information"' (GEOSS, 2005). There is some overlap between the principles and the actors for GEOSS, the Committee on Data for Science and Technology (CODATA) and the International Council for Science (ICSU), and their scientific data cultures are very similar at a macro scale. Concurrently, the Infrastructure for Spatial Information in the European Community (INSPIRE) Directive (2007), although it has not met up with its ideal of developing fully operational spatial data infrastructures for all EU member states, it deserves much credit as it standardised and made accessible key national framework datasets and spearheaded the creation of spatial data portals. INSPIRE aligns with the EU Public Sector Information (PSI) Re-Use Directives (EC, 2003), and EU digital strategies including open data.

OPEN SPATIAL DATA

Spatial data framed as accessible may not be considered as open according to open data definitions, especially if access licences are not deemed liberal enough or the formats are proprietary. The corollary is that these same data may be deemed more open than open data at a systemic level if framed in the context of spatial data infrastructures (SDIs). SDIs, for example, include ensuring: interoperability; open standards, specifications and architecture; institutional and financial contingencies; data policies; metadata standards; data models and scientific and open methodologies; data management; security; data quality and authority (Minnesota Governor's Council on Geographic Information, 1997; GSDI, 2000; NRCan, 2001). Spatial data, as just discussed, but not necessarily spatial media, might also best be situated within the discursive regime or the data culture of science (social and natural), geomatics, GIS, GIScience and remote sensing. In addition, accessible spatial data in a data preservation context might be associated with concepts such as sustainable management, data life cycle, spatial data as state records, preservation formats, authenticity, long-term plans, data sharing and deposit agreements, and accompanying documentation and methodological guides (DRI, 2011; Lauriault et al., 2015). Archived data, although made accessible, may or may not be open data. If the discussion centres around spatial data quality (Guptill and Morisson, 1995) in lieu of the qualities of the data, as is the case for open data, then concepts such as bias, representation, spatial extent, limitations, methodological rigour and models are more likely to occur. Open spatial data can also be framed within the discourses of openness, transparency and participation centred on ideas of freedom of information and open government (see Chapter 18).

The concept of open spatial data is often conflated with the following:

- Access to spatially enabled *volunteered social media data* (see Chapters 5 and 6) whereby these data are public but are not intended to be open or to be re-used by third parties, even though privacy policies and terms of use of platform owners generally state that they do resell anonymised data. Furthermore, once the data are volunteered to the platform, data intellectual property rests with the platform's owner (Saunders et al., 2012).

- Private but *machine 'accessible' scraped metadata* with location and time attributes that are not intended to be public or open data (see Chapter 22).

- The ownership of *government data* (e.g. census, demographics) and *administrative data* (e.g. unemployment, income or education) that may or may not be openly licensed but remain the copyright of the state or the Crown; in other words data are lent to citizens but are not owned by the citizens or private sector entities who access and re-use them, unless deemed to be in the public domain as is the case in the USA.

- *Crowdsourced, VGI* and *citizen science data* (see Chapter 12), which are created in myriad ways, are licensed in an ad hoc fashion or not at all, even though cultural data practices make them 'seem' open. These may also be disseminated in systems and platforms that are vulnerable to scrapers.

- *Linked spatial data*, which are spatial data engineered in such a way that they are more accessible and discoverable but once accessed may not necessarily be open (see Chapter 13).

- *Indicators and dashboards*, which make data visible, tangible and accessible but their underlying data and algorithms may remain black-boxed even though they may be available under an open licence (see Chapters 7 and 11).

- *Research data*, which may be state funded but remain the intellectual property of the researcher or the institution within which they have been created; they are often unlicensed and/or not deposited for future use or might be sensitive in nature, requiring protection for privacy reasons.

- *Local and Indigenous Knowledge*, which may be public geospatial data that are contributed by communities and elders, but these are restricted for spiritual, re-ligious and tribal affiliation reasons. They may also be restricted because sites are environmentally sensitive or are hunting and fishing grounds under protection from exploitation (Scassa et al., 2015).

- Finally, the term 'open' is used by the private sector, but is in fact a co-optation of the meaning that stands for any 'openly' acquired data that have been 'value added' and then transformed into proprietary data for commercial purposes (ODX, 2016).

Indeed, spatial media data, especially from social and locative media, are more usefully framed with respect to 'access to data' rather than 'open data', as they are privately owned, often held in proprietary formats and made available through licensing and APIs.

To confuse matters, the term 'open data' is erroneously attributed by its advocates as the starting point for the drive to make data widely accessible (Chapter 18), masking the fact that spatial, scientific and research data were made open and accessible much earlier (see Figure 9.1). In addition, these earlier accessible spatial data are considered to be closed data by open data enthusiasts or the OpenStreetMap community because of the nature of the licences[7] and data formats[8] or because there might be a small cost associated with their dissemination.[9] Ironically, open data is often simplistically and technologically framed as being simply about open data formats and open licences, yet one of the preferred 'open' spatial data formats is keyhole markup language (KML), which is by no means fully open, while the intellectual property of open data or open volunteered data mediated by third-party spatial media applications and platforms such as Google Maps (Saunders et al., 2012) renders open data as proprietary since it is plat-form providers and owners that claim ownership once the data are mediated. In this case, the medium creates spatial media by enabling access to open spatial data and by doing so transforms these to closed data. These 'closed' spatial media have nonetheless had a democratising effect, in terms of broadening access to the map-making pro-cess, the distribution of new kinds of knowledge, and advancing citizen-led evidence-informed policy but the democratising effect is however uneven (Haklay, 2013).

CONCLUSION

It is safe to say that there are no spatial media without spatial data. Spatial media uti-lise vast quantities of spatial data and sometimes generate new spatial data. However, while some of the spatial data that spatial media consume might be open, the spatial data that spatial media generate are generally not likewise, since these are rendered and disseminated through third-party proprietary applications and licences, unless of course these are in open-source platforms such as QGIS or R where the licence is explicitly open.

Are open spatial data the same as the data that form part of the discursive regime of access to data? To those who collect, manage and disseminate them, yes, as it is not so much the licence that matters – it is the sharing that does – and if they are openly shared and can be re-used on the internet they are considered open. Are open spatial data the same as open data? Probably not according to strict open data definitions; for example, even the EU PSI Re-Use that was responsible for opening up a treasure trove of data, was considered a closed and restrictive licensing regime because it limited commercial re-use. There are, however, many spatial data in open data portals that are open data.

Understandings of 'open', 'closed' and 'accessible' therefore vary, and how these are interpreted is associated with the discursive regimes, data cultures and the socio-technological data assemblages within which the spatial data discussed are situated. This chapter provides but a small fraction of possible narratives (e.g. excluding the library community, archives and cartographers) and it does not address the contentious issue of sustainably funding data catalogues and their associated cost (Kitchin et al., 2015). Irrespective, spatial data and spatial media, no matter how they are framed, open or accessible, cannot be considered as just a unique arrangement of objective and neutral facts: they are ubiquitous and diverse, and are the centre of one of the many overlapping constellations of socio-technical data assemblages (Kitchin and Lauriault, in press).

NOTES

1. http://www.domesdaybook.co.uk/
2. See, for example: The Atlas of Canada, 2nd edition, the World map – http://geogratis.gc.ca/api/en/nrcan-rncan/ess-sst/8c1daad5-7996-5d6a-8405-fea284b960ea.html
3. See, for example: The Atlas of Canada, 5th edition, Indian Treaties – http://geogratis.gc.ca/api/en/nrcan-rncan/ess-sst/fe5cf14e-5081-5c43-9822-a6da6ef5d326.html
4. See, for example: Ordnance Survey Ireland (OSi) Historical Mapping – http://www.osi.ie/products/professional-mapping/historical-mapping/
5. Administrative data result from the management of government administrations and their related services and programmes (e.g. budgets, number of student loan recipients); government data include those collected by the state as a function of the state (e.g. census data, geomatics data, health surveillance data); research data are data derived from state-funded R&D.
6. Data politics or data activism is broader than open data advocacy. The coalitions around the cancellation of Canada's Long-Form Census in 2010 by the Harper Conservative Government, the Evidence For Democracy Campaign, or the End the Back Log campaign are forms of data politics and activism (Lauriault, 2015; Lauriault and O'Hara, 2015). Open data is primarily about data policy in open government, not data politics or grassroots data activism.
7. For example, accessible spatial data may not be considered open if licences include clauses like the following – 'the data cannot be used for commercial purposes', 'the data are to be used for educational purposes only', 'the data cannot be used without permission' – if there are disclaimers about liability, indemnification or a stipulation that access to the data is for a specific amount of time, or if there is a statement that

access can be terminated at any time if the users frame the data in a way the state
does not agree with.
8. Open data advocates list a number of open formats; lists are inconsistent and
    often definitions lack clarity. See the Open Data Index (http://index.okfn.org/)
    and US National Archives and Records Administration (http://www.archives.gov/
    records-mgmt/policy/transfer-guidance-tables.htm) in order to see how archivists
    are managing knowledge about formats.
9. Some data require the paying of a dissemination fee, the cost of mailing the data on
    a datakey, or a membership or consortium fee to cover the cost of managing the
    data if not done so by a government body. Cost recovery on the one hand reduces
    access to data for those who cannot afford them; alternatively, if there is a lack of a
    sustainable business plan, data may not be maintained, or quality may be reduced. In
    some cases, as was the case with Ordnance Survey Ireland, data quality was preserved
    during the economic crash in Ireland as a result of cost recovery and its legal status
    as a state body (http://www.osi.ie/about/).

ACKNOWLEDGEMENTS

The research for this chapter was funded by a European Research Council Advanced
Investigator grant, The Programmable City (ERC-2012-AdG-323636).

REFERENCES

Ajah (2016) 'Fundtracker Pro'. Available at: http://ajah.ca/ (accessed 25 July 2016).
Arthur, C. and Cross, M. (2006) 'Give us back our crown jewels', *Guardian*, 9 March.
    Available at: https://www.theguardian.com/technology/2006/mar/09/education.
    epublic (accessed 25 July 2016).
Berners-Lee, T. (2006) '5 Star of Open Data'. Available at: http://5stardata.info/en/
    (accessed 25 July 2016).
Chrisman, N. (2006) *Charting the Unknown: How Computer Mapping at Harvard Became
    GIS*. Redlands, CA: Esri Press.
Coleman, D.J. and McLaughlin, J.D. (1997) 'Defining global geospatial data infrastruc-
    ture (GGDI): components, stakeholders and interfaces', *Geomatica* 52 (2): 129–43.
Dalby, D. and Scott, M. (2015) 'Ireland, accused of giving tax breaks to multinationals,
    plans an even lower rate', *NYTimes Online*, 13 October. Available at: http://www.
    nytimes.com/2015/10/14/business/international/ireland-tax-rate-breaks.html
    (accessed 25 July 2016).
Digital Repository of Ireland (DRI) (2011) Available at: http://www.dri.ie (accessed 25
    July 2016).
European Commission (EC) (2003) 'European legislation on reuse of public sector
    information'. Available at: https://ec.europa.eu/digital-single-market/en/european-
    legislation-reuse-public-sector-information (accessed 25 July 2016).
Fisher, T. and MacDonald, C. (1979) 'An overview of the Canada Geographic Information
    System (CGIS)', *Proceedings of the International Symposium on Cartography and Computing:
    Applications in Health and Environment, Auto-Carto* 4(1): 610–15. Available at: http://
    mapcontext.com/autocarto/proceedings/auto-carto-4-vol-1/pdf/an-overview-of-
    the-canada-geographic-information-system(cgis).pdf (accessed 25 July 2016).

Foucault, M. (2007) *Security, Territory, Population*. London: Palgrave Macmillan.

G8 (2013) 'G8 Open Data Charter'. Available at: https://www.gov.uk/government/publications/open-data-charter/g8-open-data-charter-and-technical-annex (accessed 25 July 2016).

Gartner (2015) *How to Adopt Open Data for Business Data and Analytics – And Why You Should*. New York: Gartner.

Global Earth Observation System of Systems (GEOSS) (2005) 'What is Geo?' Available at: http://www.earthobservations.org/wigeo.php (accessed 25 July 2016).

Global Earth Observation System of Systems (GEOSS) (2007) 'GEOSS data sharing principles'. Available at: https://www.earthobservations.org/geoss_dsp.shtml (accessed 25 July 2016).

Global Spatial Data Infrastructure (GSDI) (2000) 'Spatial data infrastructure cookbook'. Available at: http://gsdiassociation.org/index.php/publications/sdi-cookbooks.html (accessed 25 July 2016).

Guptill, S.C. and Morrison, J.L. (1995) *Elements of Spatial Data Quality*. Amsterdam: Elsevier Science.

Haklay, M. (2013) 'Neogeography and the delusion of democratisation', *Environment and Planning A*, 45 (1): 55–69.

Holland, D. (2015) 'Oblique imagery at Ordnance Survey Great Britain (OSGB)'. Available at: http://www.eurosdr.net/sites/default/files/images/inline/07_holland_etal-eurosdr_isprs-southampton2015.pdf (accessed 25 July 2016).

IATI (2016) 'International Aid Transparency Initiative Standard'. Available at: http://iatistandard.org/ (accessed 25 July 2016).

Infrastructure for Spatial Information in the European Community (INSPIRE) (2007) 'About us'. Available at: http://inspire.ec.europa.eu/index.cfm/pageid/48 (accessed 25 July 2016).

International Open Data Charter (2016) Available at http://opendatacharter.net/

International Polar Year (IPY) Canada (2010) 'Data management'. Available at: http://www.api-ipy.gc.ca/pg_IPYAPI_052-eng.html (accessed 25 July 2016).

Kitchin, R. (2014) *The Data Revolution: Big Data, Open Data, Data Infrastructures and their Consequences*. London: Sage.

Kitchin, R. and Lauriault, T.P. (in press) 'Towards critical data studies: charting and unpacking data assemblages and their work', in J. Eckert, A. Shears and J. Thatcher (eds), *Geoweb and Big Data*. Lincoln, NB: University of Nebraska Press.

Kitchin, R. and McArdle, G. (2016) 'What makes big data, big data? Exploring the ontological characteristics of 26 datasets', *Big Data and Society*, 3: 1–10.

Kitchin, R., Collins, S. and Frost, D. (2015) 'Funding models for open access digital data repositories', *Online Information Review*, 39 (5): 664–81.

Lauriault, T.P. (2012) 'Data, infrastructures and geographical imaginations', unpublished PhD Dissertation, Carleton University, Ottawa. Available at: http://www.collectionscanada.gc.ca/obj/s4/f2/dsk4/etd/MQ89050.PDF

Lauriault, T.P. (2015) 'Evidence-informed activism and data-based deliberations', Shaping Dublin: A Seminar Series on the Contemporary City by the Provisional University. Available at: http://www.slideshare.net/TraceyLauriault/evidence-informed-activism-databased-deliberations (accessed 25 July 2016).

Lauriault, T.P. and O'Hara, K. (2015) '2015 Canadian election platforms: long-form census, open data, open government, transparency and evidence-based policy and science', *SSRN*. Available at: http://papers.ssrn.com/sol3/papers.cfm?abstract_id=2682638 (accessed 25 July 2016).

Lauriault, T.P., Hackett, Y. and Kennedy, E. (2015) 'Geospatial data preservation primer', Canadian Geospatial Data Infrastructure, Information Product 36e. Available at: http://geoscan.nrcan.gc.ca/starweb/geoscan/servlet.starweb?path=geoscan/fulle. web&search1=R=296299 (accessed 25 July 2016).

McKinsey Global Institute (MGI) (2013) 'Open data: unlocking innovation and performance with liquid information'. Available at: http://www.mckinsey.com/ business-functions/business-technology/our-insights/open-data-unlocking- innovation-and-performance-with-liquid-information (accessed 25 July 2016).

McNerney, L. (2016) 'What the future holds for GIS: a look at the evolution of the role GIS plays in Irish society and how OSi are adapting its service offerings to meet continued demand in this space', Ordnance Survey Ireland blog. Available at: http:// www.osi.ie/blog/future-of-gis/ (accessed 25 July 2016).

Minnesota Governor's Council on Geographic Information (1997) 'Laying the foun- dation for a geographic data clearinghouse'. Available at: http://server.admin.state. mn.us/pdf/gisclear.pdf (accessed 25 July 2016).

Mueffelmann, K. (2015) 'Uber's privacy woes should serve as a cautionary tale for all companies', *Wired*. Available at: http://www.wired.com/insights/2015/01/uber- privacy-woes-cautionary-tale/ (accessed 25 July 2016).

National Academy of Sciences (NAS) (2002) 'Down to earth: geographic information for sustainable development in Africa', *National Academies Press*. Available at: http:// www.nap.edu/catalog/10455/down-to-earth-geographical-information-for-sus- tainable-development-in-africa (accessed 25 July 2016).

National Research Council (NRC) (1993) 'Toward a coordinated spatial data infra- structure for the nation'. Available at: http://www.nap.edu/catalog/2105/toward-a- coordinated-spatial-data-infrastructure-for-the-nation (accessed 25 July 2016).

Natural Resources Canada (NRCan) (2001) *Canadian Geospatial Data Infrastructure Target Vision*, prepared by the CGDI Architecture Working Group Version: 1 March 27. 2001. Available at: http://geoscan.nrcan.gc.ca/starweb/geoscan/servlet.starweb?path=geos can/fulle.web&search1=R=288842 (accessed 25 July 2016).

Natural Resources Canada (2015) *National Atlas of Canada Online 1999*. Available at: http:// www.nrcan.gc.ca/earth-sciences/geography/atlas-canada/about-atlas-canada/16890 (accessed 25 July 2016).

Open Data 500 (2016) 'Open Data 500 companies'. Available at: http://www.opendata 500.com/us/list/ (accessed 25 July 2016).

Open Data Barometer (2016) 'The Open Data Barometer'. Available at: http://opendata barometer.org/ (accessed 25 July 2016).

Open Data Exchange (ODX) (2016) 'About us'. Available at: http://codx.ca/about-us/ (accessed 25 July 2016).

Open Government Partnership (OGP) (2011) 'Open Government declaration'. Available at: http://www.opengovpartnership.org/about/open-government-dec laration (accessed 25 July 2016).

Open Knowledge Foundation (2005) 'Open Definition history'. Available at: http:// opendefinition.org/history/ (accessed 25 July 2016).

Open Knowledge Foundation (2014) 'Open Data index'. Available at: http://index. okfn.org/ (accessed 25 July 2016).

Open Knowledge Foundation (2016) 'Open definition'. Available at: http://opendefini tion.org/ (accessed 25 July 2016).

Ordnance Survey Ireland (OSi) (2014) 'Data concepts and data model overview'. Available at: http://www.osi.ie/wp-content/uploads/2015/04/Prime2-V-2.pdf (accessed 25 July 2016).

Public Sector Digest (PSD) (2015) 'Open Cities Index 2015 report'. Available at: https://www.publicsectordigest.com/Articles/view/1547 (accessed 25 July 2016).

Riofrio, M. (2016) 'How Ford's autonomous test vehicles make 3D LiDAR maps of the world around them: as Ford triples its autonomous research fleet, it's also getting new toys, including a LiDAR small enough to build into a sideview mirror', *PCWorld*. Available at: http://www.pcworld.com/article/3020407/ces/how-fords-autonomous-test-vehicles-make-3d-lidar-maps-of-the-world-around-them.html (accessed 25 July 2016).

Robbins, M. (2015) 'Porn data: visualising fetish space', *Guardian*, 30 April. Available at: https://www.theguardian.com/science/the-lay-scientist/2015/apr/30/porn-data-visualising-fetish-space (accessed 25 July 2016).

Robinson, R. (2015) 'Two important lessons from the Ashley Madison breach', *IBM Security Intelligence*. Available at: https://securityintelligence.com/two-important-lessons-from-the-ashley-madison-breach/ (accessed 25 July 2016).

Saunders, A., Scassa, T. and Lauriault, T.P. (2012) 'Legal issues in maps built on third-party base layers', *Geomatica*, 66 (4): 279–90.

Scassa, T., Taylor, D.R.F. and Lauriault, T.P. (2015) 'Cybercartography and traditional knowledge: responding to legal and ethical challenges', in R.D.F. Taylor and T. Lauriault (eds), *Developments in the Theory and Practice of Cybercartography: Applications and Indigenous Mapping*. Amsterdam: Elsevier. pp. 279–97.

Scharl, A. and Tochtermann, K. (2007) *The Geospatial Web: How Geobrowsers, Social Software and the Web 2.0 are Shaping the Network Society*. Berlin: Springer.

Scientific Committee on Antarctic Research (SCAR) (2016) 'Data and products'. Available at: http://www.scar.org/scadm/scadm-about/25-data-products (accessed 25 July 2016).

Secretariat of the Antarctic Treaty (1959) 'The Antarctic Treaty'. Available at: http://www.ats.aq/e/ats.ht (accessed 25 July 2016).

Taylor, D.R.F. and Lauriault, T.P. (eds) (2014) *Developments in the Theory and Practice of Cybercartography: Applications and Indigenous Mapping*, 2nd edition. Amsterdam: Elsevier.

United Nations (UN) (1992) 'Agenda 21, Chapter 40: Information for decision making'. Available at: http://www.un.org/earthwatch/about/docs/a21ch40.htm (accessed 25 July 2016).

William the Conqueror (commissioned) (1086) *The Domesday Book*. Available at: http://www.nationalarchives.gov.uk/domesday/ (accessed 25 July 2016).

Wilson, C. and O'Neil, B. (1999) 'GeoGratis: a Canadian geospatial data infrastructure component that visualises and delivers free geospatial data sets', unpublished paper, Natural Resources Canada. Available at: http://datalibre.ca/2013/09/23/unpublished-unpublished-paper-geogratis-a-canadian-geospatial-data-infrastructure-component-that-visualises-and-delivers-free-geospatial-data-sets/ (accessed 25 July 2016).

Winichakul, T. (1994) *Siam Mapped: A History of the Geo-Body of a Nation*. Honolulu, HI: University of Hawaii Press.

World Wide Web Consortium (W3C) (2016) 'About us'. Available at: https://www.w3.org/Consortium/ (accessed 25 July 2016).

10
GEOSPATIAL BIG DATA

DANIEL SUI

INTRODUCTION

The geospatial data landscape has changed significantly during the past decade due to advances in several key technologies related to global positioning systems (GPS), cloud computing, smartphones, ubiquitous computing (ubicomp), Web 2.0, social media, drones and the internet of things (IoT) (Hersent et al., 2012; Sui, 2013; Lee and Kang, 2015). The scope of what is considered as geospatial data has also drastically expanded. Further, instead of the traditional distinction of hardware and software, we have witnessed the emergence of 'everyware' (Greenfield, 2006) as ubicomp replaces the traditional mainframe and desktop computers to become the dominant paradigm for computing. The future world will be full of everyware – where people and objects are connected via distributed computing and unconstrained by geographical contexts, many of which will constitute spatial media. The precise meaning of 'spatial media' varies significantly in the literature (Crampton, 2014; Leszczynski and Elwood, 2015). In this chapter, it is used in an inclusive sense, referring to the converging technologies that deal with spatial, geographical or locational information implicit or explicitly. These spatial media mean we are moving towards the convergence of a 'virtually enhanced physical reality and physically persistent virtual space' (Smart et al., 2007).

One of the defining characteristics of this emerging computing environment we interact with on a daily basis is context-awareness, which generally refers to the capabilities of either mobile or embedded systems which can sense their physical environment, and adapt accordingly. This new computing environment has contributed to the explosive growth of geospatial big data. For geospatial data, context includes three essential elements: (1) where you are (location); (2) who you are with (identity); and (3) what resources are nearby (potential). 'Context' is a broad term that includes the proximity to people, devices, lighting, noise level, network availability and even social situations, such as whether you are with your family or a friend from school or a colleague from work. Location is only part of the contextual information, but location alone does not necessarily capture things of interest that are mobile or changing. In the age of big data, location has assumed a more important role in shaping everything we do on a daily basis (Gordon and de Souza e Silva, 2011). This chapter first reviews different types of geospatial big data, which is followed by a description of the major players in their production and use. Then, the chapter introduces some of the paradoxes and contradictions associated with geospatial big data, followed by a call for a more critical perspective to address some contemporary geospatial big data issues.

VARIETIES OF GEOSPATIAL DATA

The proliferation of ubiquitous computing contributes to what is popularly known as the big data deluge (The White House, 2014). Increasingly data carry spatial and temporal tags. Consequently there has been a dramatic increase in the varieties of geospatial big data in recent years. Among the multiple Vs (volume, velocity, veracity, viscosity, vinculation, virality, etc.) of big data (Shekhar et al., 2012; Kitchin and McArdle, 2016), variety is of particular concern for the geospatial big data deluge.

Until recently, we have tended to have a rather narrow definition of what is considered geographic data or information, which is often heavily influenced by the legacy of cartography and the surveying/mapping tradition. However, rapid advances in a plethora of technologies have radically transformed how geographic data are collected, stored, disseminated, analysed, visualised and used. This trend is best reflected in Google's mantra that 'Google Maps = Google in Maps'. The insertion of an 'in' between Google and Maps perhaps signifies one of the most fundamental changes in the history of human mapping efforts – a shift from managing geospatial data to managing all data geospatially. Today, users search through Google Maps not only for traditional spatial/map information, but also for almost any kind of digital information (such as Wikipedia entries, Flickr photos, YouTube videos and Facebook/Twitter postings) as long as they are geotagged. In contrast to the traditional top–down authoritative process of geographic data production by government agencies, both the private sector and citizens have played increasingly important roles in producing geographic data of all kinds through a bottom–up crowdsourcing process (Sui et al., 2012). As a result, we now have a great increase in the variety of geocoded data on a daily basis from molecular to global scales covering most of the things we can think of on or near the Earth's surface.

Although it is a challenging task to estimate the precise volume of geospatial data generated, we can safely say that geospatial data are an important part of the big data torrent. Of course, due to rapid technological advances, what is considered big or small is always a moving target. In the McKinsey Big Data Report (Manyika et al., 2011), personal location data were singled out as one of the five primary big data streams. With approximately 600 billion transactions per day, various mobile devices are creating approximately one petabyte (10^{15} bytes) of data per year globally. Personal location data alone in 2011 were a $100 billion business for service providers and

$700 billion to end-users (Manyika et al., 2011). The other four streams of big data identified by the McKinsey Institute – healthcare, public-sector administration, retail and manufacturing – also have a significant amount of data either geocoded or geotagged. So geospatial data are not only an important component of big data, but are actually, to a large extent, big data themselves. For the geospatial community, big data present not only bigger opportunities for the business community, but also new challenges for the scientific and scholarly communities to conduct ground-breaking studies related to people (at both individual and collective levels) and environment (from local to global scale).

In fact, the geospatial community has been tackling big data issues before 'big data' became a buzzword. Geospatial technologies have been at the forefront of big data challenges, primarily due to the large volumes of raster (remote-sensing imagery) and vector data (detailed property surveys) that need to be stored and managed. Back in 1997 when Microsoft Research initiated a pilot project to demonstrate database scalability, they used aerial imagery as the primary data. The TerraServer Microsoft developed at the time remains functional and is in use today. It sets the standard and the protocol for contemporary remote-sensing image-serving sites such as OpenTopography.org (LiDAR data). There are, however, a few important emerging trends that have appeared in the past several years that will further enrich the varieties of geospatial big data:

- Advances in indoor navigation technologies will accumulate more indoor human and object mobility data within the next five years, which will in turn enable us to understand human spatial behaviour at a much greater spatial and temporal granularity. Currently, geographic information systems (GIS) only map the outdoors where we spend 15% of our time.

- The growing use of drones will enable us to create an almost seamless data collection system from in situ/ground up to outer space with different spatial and temporal resolutions.

- The anticipated growth of the sharing economy and location-based social media worldwide in the next decade will not only create a new demand for real-time data but also produce a new wave of geospatial big data.

- The further expansion of the IoT will enable more objects, devices and animals to be connected together.

- As a result of the advances of the human genome project, more geocoded DNA-sequencing data will be available to the public, leading to the development of geneGIS and spatial/geographical analysis starting from the genetic level.

KEY PLAYERS IN THE GEOSPATIAL BIG DATA ASSEMBLAGE

According to Kitchin (2014), a big data assemblage is a socio-technical system composed of many apparatuses and elements that are intimately connected. In the case of geospatial big data, its assemblage includes key government agencies, industry, non-government organisations (NGOs), academic researchers and citizens from all walks of life. Although these players work together in complex, negotiated and often unanticipated ways, each plays a relatively distinctive role and these roles need to be critically reflected upon.

For example, we need to be aware that the key technologies behind the geospatial big data deluge, such as GPS, the internet, remote sensing, and to some extent GIS, have military origins, and none of these technologies was originally designed for civilian applications. They were initially developed by a suite of government–military–industrial complexes. Certain key components of these technologies (such as GPS NAVISTAR satellites or ground control stations) are still tightly controlled by the government/military. For the past three decades in general, and the last decade in particular, we have witnessed the growing role played by industry and citizens in the collection and dissemination of geospatial big data. The roles of government, industry, researchers and citizens may be changing in unexpected ways, which may lead to the transformation of all aspects of governments, businesses, education and scientific research. For example, according to Gantz and Reinsel (2011), the volume of the world's data will increase 50 times more by 2020. To store these data, we will need 75 times more IT-related infrastructure in general and ten times more servers to handle the new data. Individuals will most likely not be able to possess this type of storage capacity. The government and increasingly the private sector are gaining custody of more and more big data. For geospatial big data, government agencies still control authoritative (head) data such as geographic framework data, ground control stations, etc., but the private sector is gaining control, almost exclusively, on the crowdsourced/non-authoritative (tail) data.

Although geospatial big data are still in the hands of the government, and increasingly the private sector, we have seen the growing involvement of citizens in the production of new geospatial data in the spirit of open science, as demonstrated by the phenomenal success of OpenStreetMap (Sui et al., 2012). The integration of social media, ubicomp and urban informatics has engaged and brought citizens together in a number of worthy causes. Of particular note is the area of emergency management and disaster relief. Ushahidi, InRelief, Sahana and Crisis Commons have played crucial roles in various disaster relief efforts, relying on geospatial big data as their primary data source. These new developments, in turn, have further encouraged governments to be more open and transparent and persuaded them to inform policy changes such as the US policy towards Iran's nuclear sites (Aday and Livingston, 2009) or the Chinese government's changing policy on monitoring PM2.5 (fine particle pollution) in its major cities (Kay et al., 2015). More geocoded data are now available online (e.g. geo.data.gov, nationaldata.gov.cn) and new government-supported platforms are being developed to facilitate these developments (www.geoplatform.gov, www.tianditu.com; see Chapters 9 and 18), yet the private sector still dominates the cloud computing platforms and social media sites.

Coupled with industry-initiated efforts for open geospatial data standards, these new developments will greatly facilitate interoperability among the rather heterogeneous computing platforms. Perhaps more importantly for us as individual researchers and scholars, websites such as Open Scholar (openscholar.harvard.edu), Wikiversity, Citizendium and Scholarpedia will further facilitate openness, sharing and collaboration among researchers and scholars, by following an open science model. Furthermore, an interdisciplinary group of scholars has proposed several intriguing conceptual frameworks for us to understand the broader implications of geospatial big data (Kitchin and Dodge, 2011; Farman, 2012; de Souza e Silva and Frith, 2012), which may serve as a guiding framework for understanding their complexities. We also need to think critically about the complexity and multidimensionality of geospatial big data (boyd and Crawford, 2012; Barnes, 2013; Gorman, 2013).

CHALLENGES OF GEOSPATIAL BIG DATA: TOWARDS A CRITICAL PERSPECTIVE

Geospatial big data pose considerable technical challenges as the quality of geospatial big data is often problematic – they often lack a sampling scheme, quality control, metadata and generalisability (Goodchild, 2013). Also, how do we go about analysing and synthesising the variety of data when interoperability, a common ontology, or semantic compatibility is absent (Sui and Zhao, 2015)? Significant progress has been made to address these technical challenges in recent years, but it remains a thorny issue. What is more problematic, perhaps more challenging, is to understand a set of paradoxes and contradictions underlying the geospatial big data deluge. This requires us to move beyond technical issues and engage in a broader conversation and adopt a more critical perspective (Haklay, 2013).

TRACKING EVERYTHING VS. LOCATION SPOOFING

The abundance of geospatial big data has enabled us to track everything on or near the surface of the Earth. But for a variety of different reasons, more and more people are wary of revealing their locational information (see Chapter 22). Not surprisingly, advances have been made on how to spoof one's location – an action to masquerade as being elsewhere to confuse others – and there are plenty of tools available to do this quite easily. The challenge is: how to differentiate between true, legitimate locational information vs. false, spoofed locational information (Zhao and Sui, 2014). With whom should we share precise locational information, when to spoof our location, and for what purposes? As demonstrated in the 2009 worldwide support for the post-election protest in Iran (Sui, 2009), location spoofing was an effective weapon to counter government surveillance. We also need to be keenly aware of the contradiction that, as it is getting easier to track people and objects using these cutting-edge location-sensing technologies, it is also simultaneously getting easier to spoof one's location with spoofing apps (Zhao and Sui, 2014). Perhaps the biggest irony in the age of geospatial big data is, sadly, the disappearance of Malaysian Airline Flight 270 on 8 March 2014. As of today, the international community still has no clue as to its location. Even more problematic was the silence of the research community in general and the geospatial community in particular about this tragic event.

OPEN DATA VS. THE DEEP WEB/DARKNET

To realise the full potential of geospatial big data, there has been a strong push by a number of key actors to embrace the value of open science. Elsewhere, I have argued for an open GIS paradigm that has at least the following eight dimensions – open hardware, software, data, research, standards, funding, publication and education (Sui, 2014). Although open science is frequently touted as being common, there are still seemingly insurmountable administrative, legal, business and cultural barriers to being able to fully implement the vision of open science. Due to concerns over national security and the increasing competition in the data–driven global economy, some scholars for example have argued for data sovereignty, which refers to the concept that digital data are subject to the laws of the country in which they are located/produced (Auslander, 2015). As a result, more and more data are residing in the deep web – The portion of the web not indexed by any of the standard search engines (Sui et al., 2015). Further, with the initial release of The Onion Router (TOR) in 2002, a piece of free software for enabling anonymous communication, an increasing amount of traffic on the deep web has been flowing over encrypted network routers – in the darknet. Consequently, while enormous efforts are being made to embrace geospatial big data with the spirit of an open science, we are also simultaneously witnessing more data traffic flowing in the deep web and the darknet, which are inaccessible to most users and many policy-makers. Although it is technically impossible to estimate accurately the size of the deep web, it is telling that as early as 2001 the deep web was already several orders of magnitude larger than the massive surface web (Sui et al., 2015). The occupants of this websphere are hackers, hoaxers, whistleblowers, political extremists, vigilantes, drug dealers and spies. Their motivation for staying anonymous varies, from government surveillance to protecting personal privacy and to hiding illegal activities. We will need to resolve the paradoxes of open data vs. the growing traffic over the deep web/darknet in the context of geospatial big data.

BIG DATA DELUGE VS. THE FIRST LAW OF GEOSPATIAL BIG DATA

Along with the growth of this ever-expanding digital universe filled with geospatial big data, the world (people, manufactured objects and/or other things, and environment) is increasingly being recorded, referenced and connected by vast digital networks. Again, paradoxically, as some parts of the world are flooded by big data and people are increasingly connected, we must also be keenly aware that this world remains a deeply divided one – physically, digitally and socially. In the context of geographic information (and to some extent other types of data as well), ironically, information is usually the least available where it is most needed. It is perhaps most appropriate to call this the first law of geospatial big data. We have witnessed this paradox unfolding with respect to the Syria refuge crisis, the global diffusion of ebola virus and the brutal acts committed by various terrorist groups in different parts of the world. Undoubtedly, how to deal with big data in a shrinking and stratified world remains a major challenge in the geospatial big data era. To understand these growing uneven information and communication geographies, we must move away from the traditional, linear conceptualisation of a digital divide, concerned primarily with physical access to computers and the internet. Instead, we must consider the multiple divides within cyberspace (or digital apartheid) by taking into account the hybrid, scattered, ordered and individualised nature of cyberspaces (Graham, 2011). Indeed, multiple hidden social and political factors are at play in determining what is or is not available online. The first law of geospatial big data produces (and is also simultaneously produced by) the digital divide. To make this world a sustainable one, an essential step is to narrow the divide and eventually defeat the first law of geospatial big data.

GREEN VS. GREY/E-WASTE

Last, but not least, there are concerns over the long-term environmental impacts of geospatial big data. In this hybrid world dominated by O2O (online to offline), will the flows of geospatial big data automatically contribute to the emergence of a smart planet at the global level? Or will the deluge of geospatial big data lead to more ubiquitous consumption, thus more material and energy consumption that will accelerate resource depletion and environmental devastation? Or will the abundant geospatial big data further promote citizen science and better environmental monitoring and product tracking through their entire life cycle, thus helping in the effort to save our planet? Are we closer to or further away from the goals of sustainability in the age of geospatial big data? How can geospatial big data be deployed to advance the cause for the environment? Will the geospatial big data we collect from the sensor networks for animals, plants and physical elements result in new types of environmental rights? Furthermore, scientists and artists alike have been concerned about the e-wastes generated by the big data-based digital economy, the energy consumption of the big data server farms with cloud computing storage, and long-term effects on the environment. Instead of the naïve view that the digital economy is green, we need to conduct more critical analyses on the environmental consequences of geospatial big data.

SUMMARY AND CONCLUSIONS

The goal of this chapter has been to present an overview about the variety, players and challenges of the emerging geospatial big data. The computing environment has undergone a fundamental shift during the past two decades as computing has gradually embedded itself into our daily environment. For the first time in human history, we have the capability of tracking the location of individuals and objects in real time. The growth of ubicomp has led to a new sentient environment in which the virtual and physical world, digital bits and atoms, people and objects are linked. This computing environment also contributes to the spatial big data deluge. The new spatial big data deluge poses formidable technical challenges, but its social/political challenges are also perplexing. Moving forward, there are multiple challenges along computational, theoretical, social, political, legal and environmental fronts. To better address these challenges, we need to take a more critical perspective and examine and decipher the geospatial big data assemblage. The positive impacts of geospatial big data on society and economy are enormous; however, geospatial big data have also created a new form of *terra incognita* calling for more critical inquiry.

REFERENCES

Aday, S. and Livingston, S. (2009) 'NGOs as intelligence agencies: the empowerment of transnational advocacy networks and the media by commercial remote sensing in the case of the Iranian nuclear program', *Geoforum*, 40 (4): 514–22.

Auslander, D. (2015) 'Data sovereignty and the cloud: how do we fully control data amidst the cloud sprawl?', *Cloud Computing*. Available at: http://www.cloudcomputing-news. net/news/2015/sep/09/data-sovereignty-and-cloud-how-do-we-fully-control-data-amidst-cloud-sprawl (accessed 25 July 2016).

Barnes, T.J. (2013) 'Big data, little history', *Dialogue in Human Geography*, 3 (3): 297–302.

boyd, d. and Crawford, K. (2012) 'Critical questions for big data', *Information, Communication and Society*, 15 (5): 622–79.

Crampton, J. (2014) 'New spatial media', *Open Geography*. Available at: https://opengeography.wordpress.com/2014/06/06/new-spatial-media/ (accessed 25 July 2016).

de Souza e Silva, A. and Frith, J. (2012) *Mobile Interfaces in Public Spaces: Locational Privacy, Control, and Urban Sociability*. New York: Routledge.

Farman, J. (2012) *Mobile Interface Theory: Embodied Space and Locative Media*. New York: Routledge.

Gantz, J. and Reinsel, D. (2011) 'Extracting value from chaos'. Available at: http://www. emc.com/collateral/analyst-reports/idc-extracting-value-from-chaos-ar.pdf (accessed 25 July 2016).

Goodchild, M.F. (2013) 'The quality of big (geo)data', *Dialogue in Human Geography*, 3 (3): 280-4.

Gordon, E. and de Souza e Silva, A. (2011) *Net Locality: Why Location Matters in a Networked World*. Malden, MA: Wiley-Blackwell.

Gorman, S.P. (2013) 'The danger of a big data episteme and the need to evolve geographic information systems', *Dialogue in Human Geography*, 3 (3): 285–91.

Graham, M. (2011) 'Time machines and virtual portals: the spatialities of the digital divide', *Progress in Development Studies*, 11 (3): 211–27.

Greenfield, A. (2006) *Everyware: The Dawning Age of Ubiquitous Computing*. San Francisco, CA: New Riders.

Haklay, M. (2013) 'Neogeography and the delusion of democratization', *Environment and Planning A*, 45 (1): 55–69.

Hersent, O., Boswarthick, D. and Elloumi, O. (2012) *The Internet of Things: Key Applications and Protocols*. New York: John Wiley and Sons.

Kay, S., Zhao, B. and Sui, D.Z. (2015) 'Can social media clear the air? A case study of the air pollution problem in Beijing', *The Professional Geographer*, 67 (3): 351–63.

Kitchin, R. (2014) *The Data Revolution: Big Data, Open Data, Data Infrastructures and their Consequences.* London: Sage.

Kitchin, R. and Dodge, M. (2011) *Code/Space: Software and Everyday Life*. Cambridge, MA: MIT Press.

Kitchin, R. and McArdle, G. (2016) 'What makes big data, big data? Exploring the onto-logical characteristics of 26 datasets', *Big Data and Society*, 3: 1–10.

Lee, J.G. and Kang, M. (2015) 'Geospatial big data: challenges and opportunities', *Big Data Research*, 2 (2): 74–81.

Leszczynski, A. (2015) 'Spatial big data and anxieties of control', *Environment and Planning D: Society and Space*, 33 (6): 965–84.

Leszczynski, A. and Elwood, S. (2015) 'Feminist geographies of new spatial media', *The Canadian Geographer*, 59 (1): 12–28.

Manyika, J., Chui, M., Brown, B., Bughin, J., Dobbs, R., Roxburgh, C. and Byers, H.A. (2011) *Big Data: The Next Frontier for Innovation, Competition, and Productivity*. Mckinsey. Available at: http://www.mckinsey.com/Insights/MGI/Research/Technology_and_Innovation/Big_data_The_next_frontier_for_innovation (accessed 25 July 2016).

Shekhar, S., Gunturi, V., Evans, M.R. and Yang, K.S. (2012) 'Spatial big-data chal-lenges intersecting mobility and cloud computing', *Proceedings of the Eleventh ACM International Workshop on Data Engineering for Wireless and Mobile Access*. New York: ACM. pp. 1–6.

Smart, J.M., Cascio, J. and Paffendorf, J. (2007) 'Metaverse roadmap overview'. Available at: http://www.metaverseroadmap.org/overview/ (accessed 25 July 2016).

Sui, D.Z. (2009) 'Twitter's new twist on geography', *GeoWorld*, September: 18–20.

Sui, D.Z. (2013) 'Ubiquitous computing, spatial big data, and open GeoComputation', in R. Abrahart, S. Openshaw and L. See (eds), *GeoComputation*, 2nd edition. London: CRC Press. pp. 375–95.

Sui, D.Z. (2014) 'Opportunities and impediments in open GIS', *Transactions in GIS*, 18 (1): 1–24.

Sui, D.Z. and Zhao, B. (2015) 'GIS as media: mediated geographies and geographies of media through the geoweb', in S. Mains, J. Cupples and C. Lukinbeal (eds), *Geographies of Media/Mediated Geographies*. Berlin: Springer. pp. 191–208.

Sui, D.Z., Elwood, S. and Goodchild, M.F. (eds) (2012) *Crowdsourcing Geographic Knowledge: Volunteered Geographic Information in Theory and Practice*. Berlin: Springer.

Sui, D.Z., Caverlee, J. and Rudesill, D. (2015) *The Deep Web and the Darknet: A Look Inside the Internet's Massive Black Box*. Washington, DC: Woodrow Wilson International Center for Scholars.

The White House (2014) *Big Data: Seizing Opportunities, Preserving Values*. Washington, DC: Executive Office of the President. Available at: http://www.whitehouse.gov/issues/technology/big-data-review (accessed 25 July 2016).

Zhao, B. and Sui, D.Z. (2014) 'True lies in big data: detecting location spoofing in social media', *Proceedings of 2014 Big Data Urban Informatics Conference*, Chicago, IL, 20-22 August.

11

INDICATORS, BENCHMARKING AND URBAN INFORMATICS

ROB KITCHIN, TRACEY P. LAURIAULT AND GAVIN McARDLE

INTRODUCTION

As introduced in Chapter 7, urban dashboards have begun to proliferate in recent years as a form of spatial media for communicating information about cities. These dashboards use a variety of graphing and mapping visualisations to display both static and real-time data concerning all aspects of urban living, including economy, service provision, health, environment, sustainability, quality of life, transport and infrastructure, derived from a number of sources. As such, dashboards constitute a form of urban informatics – that is, they visualise and model spatially referenced data to produce information about how cities are performing, how they work, and how their various systems might be improved (Foth, 2009; Burrows and Beer, 2013). Related forms of urban informatics include geographic information systems (GIS) (Chapter 2), digital mapping (Chapter 3), geodesign (Chapter 8), city operating systems, and digital kiosks on city streets that provide hyperlinked, touchscreen browsers for discovering travel guide information about a city, such as points of interest, how to get to them, local history and so on. As well as providing information for citizens and visitors, the utility of urban informatics is its facilitation of new forms of operational and policy governance for cities.

This chapter considers the technical and political aspects of the spatially referenced data that underpin the urban informatics enacted by city dashboards. It starts by detailing the characteristics of urban indicator data and the process of city benchmarking. City dashboards often include other kinds of data, such as real-time data from various sensors and scanners mostly relating to transport (e.g. bike usage, bus location, road travel speeds) and environment (e.g. weather, pollution, water levels), as well as social media (e.g. maps of Twitter and Instagram activity) and crowdsourced data (e.g. the location of potholes or graffiti); however, given these data are discussed in other chapters we concentrate on indicator data. Next we provide a critical examination of indicator data and city benchmarking, setting out concerns relating to data selection and formulation, data quality and data analysis. This is followed by an exploration of urban informatics as data assemblages; that is, as dense amalgams of actors, actants and socio-spatial processes that work together in complex and negotiated ways to produce an urban information system (Kitchin, 2014). In the final section, we detail issues that require future attention, suggesting the need for greater reflection on the data underpinning urban informatics and the detailed mapping-out of data assemblages. The argument we put forward can equally be applied to other kinds of urban data or urban informatics, such as those produced by GIS, street kiosks and urban control rooms.

URBAN INDICATORS AND CITY BENCHMARKING

Indicators are recurrent quantified measures that can be tracked over time to provide information about stasis and change with respect to a particular phenomenon (Godin, 2003). From the early 20th century a variety of social, economic and environmental indicators have been developed to track how nations and cities are performing. From the early 1990s urban indicator projects have proliferated as a result of, on the one hand, the sustainability agenda arising from the 1992 United Nations Conference on Environment and Development (UNCED) in Rio de Janeiro, which mandated cities to track indicators related to sustainable development, and on the other the rise of urban managerialism and the desire to reform the public sector management of city services by tracking performance and implementing evidence-based decision-making. Consequently, many cities around the world now routinely generate and collate suites of indicators, using them to track and trace performance, guide policy formulation, and inform how cities are governed and regulated.

Typically, there are two main types of indicators used to monitor cities. Single indicators consist of a measurement or a statistic related to a particular phenomenon, such as the unemployment rate. The most desirable single indicators are well defined and unambiguous, and have strong representativeness (they measure what they declare themselves to measure). In some cases, indicators are indirect in nature as the phenomenon of interest is intangible or not directly observable, such as the number of patent applications being used as proxy for innovation (Gruppa and Mogee, 2004). Composite indicators combine several single measures using a system of weights or statistics to create a new derived measure, such as a deprivation index that combines several indicators such as household income, employment status, welfare and health status, and access to services into a single overall score (Maclaren, 1996). Composite indicators recognise that different phenomena are interrelated and multidimensional, and that no one indicator can reveal the extent or complexities underpinning an issue such as deprivation.

These kinds of indicator data are rarely produced by spatial media. Rather, urban indicator data are usually derived from administrative datasets and are consumed by spatial media. That is, indicator data form key base data for city dashboards, through which they are visualised in various ways to produce information and insight. In general, indicators are deployed in three ways. First, as descriptive or contextual indicators that provide insights into the present state of play and the prevailing trends with respect to a particular phenomenon. Second, as diagnostic, performance, and target indicators that provide a means to diagnose a particular issue (such as whether an issue is becoming a problem or is abating), or to assess performance such as effectiveness (whether goals are being met) and efficiency (whether getting the most output for the input) of a policy or programme or organisation (Holden, 2006). Targets can be absolute (to reach a defined level) or relative (to match the performance of another organisation/place). Third, there are predictive and conditional indicators that are thought to be useful for predicting and simulating future situations and performances and thus identifying what policy and operational changes might be required in the present to reshape the future in desired ways (Maclaren, 1996).

A common way to determine the performance of a city is to benchmark indicator data both within and across cities, comparing trends between different locales or against best practice. This is often accompanied by scorecarding in which tables of rankings and ratings, along with changes in relative position, are produced to reveal which places are doing well and which are underperforming (Gruppa and Mogee, 2004). Huggins (2009) details three types of area-based benchmarking. First, performance benchmarking that compares how well a place is doing with respect to a set of prescribed indicators. Second, process benchmarking that compares the practices, structures and systems of places. Third, policy benchmarking that compares public policies that influence performance and processes with respect to outcomes and meeting prescribed expectations. Luque-Martinez and Munoz-Leiva (2005) detail that these can be benchmarked in three ways: competitive benchmarking, wherein cities are ranked and rated regardless of whether they want to participate in the process (e.g. the best or most innovative place to live); cooperative benchmarking, where cities cooperate with the benchmarker, providing necessary information but often on the basis that target cities are not direct competitors (e.g. Vital Signs[1]); and collaborative benchmarking, where several cities work together to produce standardised indicators and to share knowledge and resources (e.g. FCM Quality of Life Indicators[2]). Jones Lang LaSalle report that there are now over 150 city benchmarking initiatives that seek to compare and contrast hundreds of cities globally (Moonen and Clark, 2013). Some of these initiatives benchmark cities across a range of indicators, some focus on particular sectors such as economic performance.

A CRITICAL ASSESSMENT OF URBAN INDICATORS AND CITY BENCHMARKING

The power of urban indicators is that they seemingly provide quantitative measures that are objective, neutral and independent of external influence, traceable over time and space, verifiable and replicable, sensitive to change, and are quick and cost-effective to collect, process and update (Franceschini et al., 2007). Indicators unambiguously capture and communicate what is occurring with respect to a phenomenon and can be compared across locales. A fact is after all a fact and can be accurately measured (e.g. there are x number of people living in a city; x percentage of them are unemployed) and that fact can be calculated for all cities. But is this really the case?

Many would argue that it is not. Instead they would contend that data do not pre-exist their generation and are the product of the ideas, instruments, practices, contexts, knowledges and systems used to generate, process and analyse them (Ribes and Jackson, 2013); that there is nothing natural or inherent about data and facts. Rather they are epistemological units, made to have a representational form that enables epistemological work. Moreover, how data are ontologically defined and delimited is not simply a neutral, technical process, but also a normative, political and ethical one that is often contested (Bowker and Star, 1999). Indeed, data about the same phenomena can be measured, recorded, analysed and interpreted in numerous ways (Poovey, 1998), and how we conceive of data, how we measure them and what we do with them frame actively the nature of data (Kitchin, 2014). Urban indicator and benchmarking initiatives inherently express a normative notion about what should be measured, for what reasons and what they should tell us, and are full of values and judgements shaped by a range of views and contexts. Or as Bowker (2005) puts it, data are never raw, but always already cooked.

Moreover, indicator and benchmarking initiatives adopt an instrumental rationality that promotes a particular way of understanding the world that, on the one hand, marginalises other forms of knowing such as 'phronesis (knowledge derived from practice and deliberation) and metis (knowledge based on experience)' (Parsons, 2004: 49), and on the other hand has a number of shortcomings. Cities are highly complex, open, contingent systems consisting of a rich, multidimensional set of relations. In seeking to reduce this complexity into a set of simplified, one-dimensional measures, indicators inevitably leave out more than they capture. Moreover, they decontextualise cities from their history, political economy, the wider set of social, economic and environmental relations that frame their development, and their wider interconnections and interdependencies. In other words, a city is much more than the sum of its indicators.

With respect to city benchmarking, the method of scorecarding assumes it is both possible and desirable to measure and compare indicator data between cities. Benchmarking assumes there is a normative standard by which cities should be judged, some ideal state they are all seeking to achieve, rather than acknowledging that phenomena in different places differ from one another often for good reasons – cities have different histories and trajectories, varying political economies and varieties of capitalism, different forms of governance and policy ambitions, and varying access to resources and capacities. Why then judge them against one another in a zero-sum ranking when what we might expect or desire them to achieve is different? In other words, is city benchmarking the best way to assess the relative performance of a city? Or does it simply foster imitation designed to game city rankings rather than produce tailored urban policies conditioned to local situations and ambitions?

Further, urban indicator data and associated informatics suffer from a number of technical and methodological issues that are often overlooked or ignored. As with all data, because indicators are abstracted, generalised and approximated through their production there are always questions concerning data veracity and quality and how accurately (precision) and faithfully (fidelity) the data represent what they are meant to, especially when using samples and proxies, and how clean (error and gap free), untainted (bias free), consistent (few discrepancies) and reliable (the measurement instrument consistently producing the same quality of results) the data are (Kitchin, 2014). The level of truthfulness and trustworthiness of data also varies over time and place due to different measurement regimes and their changing nature (as technologies, practices and personnel alter) (Ribes and Jackson, 2013).

There are a number of other issues, caused by spatial resolution and aggregation and creating composite scores, that can produce a range of ecological fallacies. For example, changing the pattern and scaling of the zones used to delimit data (e.g. geographic boundaries), or how they are categorised, changes the pattern of aggregated observations, often in quite dramatic ways. This can change how a city is ranked with respect to other cities. For example, whether data refer to Dublin City Council, Dublin County or the Greater Dublin Area can markedly change the city's unemployment rate and how that rate compares with elsewhere. Similar issues exist with respect to composite indicators, whereby changing the relative weightings of indicators in the formula, or including or excluding some variables, can have a profound effect on the resulting score. Further, composite indicators need to be normalised onto common geographic, time or threshold scales; the weightings set to reflect importance and to prioritise key interrelationships; and multi-colinearity between variables or conflicts in their commensurability need to be accounted for – but often none of these tasks are performed (Böhringer and Jochem, 2007). As such, these derived data lack a sound scientific base from which to draw definitive conclusions. In other words, even at a purely technical level, indicator data need to be treated with caution rather than them being assumed to reveal the city as it really is.

This is not to say that urban indicator and benchmarking data, or the spatial media that employ them, are not useful or valuable – they quite patently are – but rather that they need to be treated with suitable caution when used, recognising and compensating for any shortcomings. The same is the case for other kinds of data used or generated by spatial media. One way to achieve this is to map out and be aware of the data assemblage that constitutes an urban informatics initiative.

UNDERSTANDING URBAN INFORMATICS AS DATA ASSEMBLAGES

In previous work we defined a data assemblage as a socio-technical system composed of many apparatuses and elements that are thoroughly entwined (Kitchin, 2014). The data assemblages that create urban indicator data include: a number of different stakeholder institutions such as city authorities, government agencies, university institutes and software vendors (each of which has differing roles and power to influence/control the system); a range of different actors including statisticians, researchers, developers, technicians and managers; an amalgam of different actants such as computers, servers, screens and databases; a suite of different legal and governmental requirements such as data protection laws, licensing regulations, policy initiatives, data standards and software protocols; access to finance such as state funding, research grants and venture capital; an understanding and entwining of different knowledges, theories and models; as well as factors relating to political economy, places, marketplaces and so on. Together these different apparatus and elements work together, sometimes harmoniously, other times in friction, to frame, shape and produce urban indicators.

The data such an assemblage produces are thus contingent, relational and contextual, the result of collaboration, negotiation and contestation, and dependent on many actors and factors. This is why the data they produce are never raw, nor are they neutral, essential and objective. Rather, the data are cooked with respect to a negotiated recipe by a team of chefs embedded within institutions that have certain aspirations and goals and operate within wider constraints (Bowker, 2005). It is also why there are: so many different types of urban indicators, generated and measured in a variety of ways; hundreds of different city indicator projects that use varying amalgams of indicators and have dashboard interfaces that differ in how the data are presented and interacted with; and there are many different city benchmarking initiatives that employ different types of scorecards using varying suites of indicators, which are given different relative weightings in importance.

It is this variation in city indicator projects that is driving standardisation initiatives that seek to create harmony across projects to enable legitimate comparison. One such initiative that has recently come to fruition is the new International Organization for Standardization (ISO) standard for city benchmarking indicators (ISO 37120: 2014). This was designed and proposed by the Global City Indicators Facility (GCIF) (cityindicators.org), a joint project of the World Bank, UN-Habitat, the World Economic Forum, the Organisation for Economic Co-operation and Development (OECD) and the Government of Canada. The ISO standard is designed to produce standardised global urban data which are verifiable and transparent, certified by an independent third party (World Council on City Data) and thus deemed trustworthy. Cities are required to produce and report up to 100 indicators with respect to 17 themes concerning city characteristics, services, infrastructure and quality of life. Here, several large supranational agencies have joined forces, working with 254 cities across 81 countries, to develop a new global standard for urban data. Such a process is inevitably highly political and involves the playing-out of a complicated set of power geometries between institutions and actors, with the design of the standard and the selection of the themes and indicators involving negotiation and diplomacy. In other words, while the standard aims to create rigour and transparency, there is little objective or neutral about its composition, nor the data that it generates and compares.

ISSUES FOR FURTHER REFLECTION

Spatial media both consume and produce data. In this chapter we have examined the nature of some of the data that spatial media such as city dashboards utilise and visualise, namely urban indicator and scorecard data. The argument we have put forward has been that, however rigorously and systematically such data are generated, they are never essential, raw, objective and neutral. They are always cooked to some recipe. Data are the products of socio-technical assemblages composed of many apparatus and elements that work together in complex, negotiated and often unanticipated ways. The production and use of data are always contingent, relational and contextual. Moreover, there are always technical and methodological issues to consider with regards to how data are processed, analysed and interpreted. What that means is that the users of urban indicator and scorecard data need to be mindful of the potential shortcomings of data and their analysis and to remember that the data do not simply show a city as it really is and how it is performing over time or vis-à-vis other cities. The message being communicated through spatial media varies as a function of the data as much as the medium.

The data within spatial media present a particular version of the city as defined by those data. Of course, the data might reveal a very compelling picture of the present state of play that seems to match knowledge derived from other observations and experience. After all, we are not suggesting that urban indicator data have no utility or validity, or provide little insight, but rather one needs to assess their veracity and fidelity to make sure that one can be confident in what the data show and how they should be interpreted. This can be quite difficult to do when most city dashboards do not provide the metadata that would enable such an assessment. Instead, the data are displayed using graphs and maps but rarely are they accompanied by information about data lineage or provenance, or data quality, calculated error rates or calibration. The user is simply expected to trust that the data are valid and reliable. The process of standardisation is one way to try to counter such concerns by ensuring that every city generates, processes and analyses its data in the same way.

Regardless, we feel that much greater reflection and assessment of urban data needs to take place to document the issues associated with different datasets. And this is the case for all kinds of data used and produced by spatial media, not just indicator and scorecard data. Moreover, it is certainly the case that we have little detailed understanding of the composition and workings of data assemblages related to urban indicator and city benchmarking initiatives, city dashboards and urban informatics more broadly. As such, there is a need for a set of empirical case studies that map out in detail the various apparatus, elements and processes that comprise an urban data assemblage. Such studies would give us a much better appreciation for the ways in which urban data are cooked and then consumed.

NOTES

1. http://www.vitalsignscanada.ca/en/home
2. Federation of Canadian Municipalities (FCM) http://www.municipaldata-donnees municipales.ca/Site/Collection/en/index.php

ACKNOWLEDGEMENTS

This chapter draws on and reworks material from Kitchin, R., Lauriault, T. and McArdle, G. (2015) 'Knowing and governing cities through urban indicators, city benchmarking and real-time dashboards', *Regional Studies, Regional Science*, 2: 1–28. The research was funded by an ERC Advanced Investigator award (ERC-2012-AdG 323636-SOFTCITY).

REFERENCES

Böhringer, C. and Jochem, P.E.P. (2007) 'Measuring the immeasurable – a survey of sustainability indices', *Ecological Economics*, 63: 1–8.

Bowker, G. (2005) *Memory Practices in the Sciences.* Cambridge, MA: MIT Press.

Bowker, G. and Star, L. (1999) *Sorting Things Out: Classification and Its Consequences.* Cambridge, MA: MIT Press.

Burrows, R. and Beer, D. (2013) 'Rethinking space: urban informatics and the socio-logical imagination', in N. Prior and K. Orton-Johnson (eds), *Digital Sociology: Critical Perspectives.* Basingstoke: Palgrave. pp. 61–78.

Foth, M. (2009) 'Preface', in M. Foth (ed.), *Handbook of Research on Urban Informatics: The Practice and Promise of the Real-Time City.* New York: IGI Global.

Franceschini, F., Galetto, M. and Maisano, D. (2007) *Management by Measurement: Designing Key Indicators and Performance Measurement Systems.* Berlin: Springer.

Godin, B. (2003) 'The emergence of S&T indicators: why did governments supplement statistics with indicators?', *Research Policy*, 32: 679–91.

Gruppa, H. and Mogee, M.E. (2004) 'Indicators for national science and technology policy: how robust are composite indicators?', *Research Policy*, 33 (9): 1373–84.

Holden, M. (2006) 'Urban indicators and the integrative ideals of cities', *Cities*, 23 (3): 170–183.

Huggins, R. (2009) 'Regional competitive intelligence: benchmarking and policy-making', *Regional Studies*, 44 (5): 639–58.

Kitchin, R. (2014) *The Data Revolution: Big Data, Open Data, Data Infrastructures and their Consequences.* London: Sage.

Luque-Martınez, T. and Munoz-Leiva, F. (2005) 'City benchmarking: a methodological proposal referring specifically to Granada', *Cities*, 22 (6): 411–23.

Maclaren, V.W. (1996) 'Urban sustainability reporting', *Journal of the American Planning Association*, 62 (2): 184–203.

Moonen, T. and Clark, G. (2013) 'The business of cities 2013: what do 150 city indexes and benchmarking studies tell us about the urban world in 2013?' Jones Lang LaSalle. Available at: http://www.jll.com/Research/jll-city-indices-november-2013.pdf (accessed 25 July 2016).

Parsons, W. (2004) 'Not just steering but weaving: relevant knowledge and the craft of building policy capacity and coherence', *Australian Journal of Public Administration*, 63 (1): 43–57.

Poovey, M. (1998) *A History of the Modern Fact: Problems of Knowledge in the Sciences of Wealth and Society.* Chicago, IL: University Chicago Press.

Ribes, D. and Jackson, S.J. (2013) 'Data bite man: the work of sustaining long-term study', in L. Gitelman (ed.), *'Raw Data' is an Oxymoron.* Cambridge, MA: MIT Press. pp. 147–66.

12

VOLUNTEERED GEOGRAPHIC INFORMATION AND CITIZEN SCIENCE

MUKI HAKLAY

INTRODUCTION

This chapter explores the related areas of Volunteered Geographic Information (VGI) and citizen science. VGI is defined as digital geographical information that is generated and shared by individuals. VGI can be viewed as the part of user-generated content that has become a major element of spatial media over the past two decades. Within VGI, geographical information is an integral part of the digital media object – for example, the coordinates embedded in the exchangeable image file format (Exif) of a picture taken with a digital camera (Goodchild, 2007). Citizen science, on the other hand, is defined by the *Oxford English Dictionary* as 'scientific work undertaken by members of the general public, often in collaboration with or under the direction of professional scientists and scientific institutions' (2014). Citizen science can also be considered as a type of user-generated content, where this content refers to scientific facts, observations or analysis.

User-generated content started early on in the World Wide Web (web), when systems such as GeoCities (launched in 1994) allowed people to link to the system over the internet and create their own websites, even though they possessed relatively limited technical skills (Brown, 2001). With further technological and interaction design advances it became possible to create content with even less technical knowledge through weblogs (blogs), images, audio (podcasts) and video-sharing websites. Around 2005, due to technical, societal and organisational changes, the process of capturing geographic information using a range of affordable devices – from global positioning system (GPS) receivers through to cameras and phones – became increasingly available to a wider group of people (see Haklay et al., 2008, for a detailed analysis of the underlying trends). Of these changes, it is worth noting one concept in particular within a societal and organisational realm: crowdsourcing. The term 'crowdsourcing' was coined by Howe (2006) to describe a process in which a large group of people are asked to perform business functions that are either difficult to automate or expensive to implement. Fundamentally, crowdsourcing allows an organisation (be it a company or a scientific research institute) to ask a large group of unremunerated or marginally remunerated people to carry out tasks for which the organisation is the prime beneficiary.

Of particular importance are the labelling of a purposeful activity that people can partake in and the use of terms such as 'volunteer', 'citizen', 'user' and 'crowd' to describe the participants. While 'volunteer' and 'citizen' are clearly loaded with meaning, 'crowd' and 'user' might seem neutral or simply descriptive. Yet, crowdsourcing has been much criticised as a potentially exploitative practice that reduces humans to automatons or machine parts (Silverman, 2014) and therefore 'crowd' is used to treat the contributors to the activity as an anonymous, faceless (and potentially expendable) group. The term 'user' within digital technology discourses has been criticised by Brenda Laurel (2001: 49–50), who observed that 'user'

> implies an unbalanced power relationship – the experts make things; everybody else is just a user. People don't like to think of themselves as users. We like to see ourselves as creative, energetic beings who put out as much as we take in.

She goes on to suggest the term

> Partner – this person has agreed to work on something together with you. The idea of being in partnership with the people purchasing your products or on your site is not only emotionally attractive; it is quite literally true.

As for 'volunteer' in VGI, this has received special attention from Sieber and Haklay (2015), who argue that the assumption of free-will volunteering, without any wish for personal gain, is not reflected in practices such as crowdsourcing where there is no explicit volunteering for a higher cause. Conversely, instead of seeing volunteering as a reason to increase the trust in the participant, it is a source of concern about their motivations. As can be expected, the 'citizen' in citizen science has also raised a lot of questions as demonstrated by Mueller et al. (2012), who argue that the use of the term 'citizen' requires the linking of public participation in science to a strong concept of democratisation and citizenship, especially when citizen science projects are linked to education. Similar sentiments are echoed in the responses to their paper by Cooper (2012) and Calabrese Barton (2012).

These are merely a few examples of a much wider literature that critiques and questions the use of these loaded terms to describe large-scale activities that have emerged in the past decade. Arguably, they are the result of the underlying tensions that are at the basis of VGI and citizen science, as either altruistic, collaborative efforts towards a common goal and a greater good on the one hand, or extracting free labour in an exploitative way where the benefits justifiably accrue to the entre-preneurs who have set up the system or have the knowledge and skills to exploit the resulting information on the other. The reality is, as expected, somewhere in between, depending on the nature of the project and its dynamics.

Beyond the discussions about the terminology and its meaning, we should notice the scale and reach of these activities, which engage millions of people across the world through data collection, information sharing and the analysis of geographic information. The remainder of this chapter examines activities that fall under the banners of VGI and citizen science, and especially the intersection between them. Throughout, the intention of volunteers in their act of participation and the issue of power between the contributor and the technical and social systems that facilitate the contribution are detailed, as well as the values that are embedded into the prac-tices of participation and information sharing.

VGI AND CITIZEN SCIENCE

Craglia et al. (2012) provide a useful framework for considering VGI and citizen science. They suggest differentiating between volunteering and geographical content. They fur-ther suggest differentiating between implicit and explicit contributions. Explicit volun-teering is when people are knowingly volunteering effort to a project, while implicit volunteering is when information is shared openly, but without people knowing how their contribution will eventually be used. For example, carrying out bird observations and reporting to a shared database is considered to be active volunteering, while the re-use of all the georeferenced images of parks that are shared on Flickr to assess the level of interest in the parks (or the cleanliness of these parks) is implicit volunteering, since the images were shared without this purpose in mind. Another helpful distinction can be made depending on whether or not the participant needs to actively and know-ingly contribute information (e.g. use an app such as WideNoise to measure the level of noise – see Becker et al., 2013) or passively share (e.g. use a phone to sense the signal from different telephone masts and share this information on OpenSignal[1]).

VGI

Within VGI, there are activities that clearly fall outside the realm of citizen science – for example, when people use their phone to 'check in' to a bar or provide a restaurant review in apps such as Yelp.[2] This is an example of volunteering as an active and explicit geographic contribution. VGI also includes contributions to Wikipedia[3] that contain place names (e.g. an article about a historical figure mentioning a place that they trav-elled to) but are not explicit geographic contributions, since the aim of the article is not geographical, although it is explicitly contributed.

Some VGI is very similar to citizen science in that it is concerned with recording geographical facts and observations. An application such as StreetBump[4] runs on a participant's phone and uses the sensors in the phone to detect bumps in the road while they are driving their car. Here, there is an explicit sensing of car movement associated with the geographic location from GPS. This is explicitly volunteered, passive and an explicit geographic contribution. OpenStreetMap[5] is another VGI example that parallels citizen science as it is concerned with recording facts about the world and measuring them accurately.

CITIZEN SCIENCE

There are a number of types of activities that constitute citizen science, six of which are discussed here: passive sensing, volunteer computing, volunteer thinking, environmental and ecological observations, participatory sensing, and community science (for a more in-depth examination, see Haklay, 2013, 2015).

1. *Passive sensing:* relies on participants in the project to provide a resource that they own (e.g. their phone) for automatic sensing. The information that is collected through the sensors is then used by scientists for analysis (e.g. StreetBump).

2. *Volunteer computing:* is a method in which participants share their unused computing resources on their personal computer, tablet or smartphone, and allow scientists to run complex computer models during the times when the device is not in use (e.g. when people use the sensor in their laptop to augment seismographic networks in the Quake-Catcher[6] project).

3. *Volunteer thinking:* uses what Shirky (2011) terms 'cognitive surplus', which is the cognitive ability of people not used in passive leisure activities, such as watching TV. In this type of project, the participants contribute their ability to recognise patterns or analyse information that will then be used in a scientific project (e.g. GeoTag-X[7] recruits volunteers to help with classification of images as part of humanitarian efforts).

4. *Environmental and ecological observation:* focuses on monitoring environmental pollution or observations of flora and fauna, through activities (e.g. bio-blitz, in which a group of volunteers studies a site thoroughly, using their phones to record and share observations).

5. *Participatory sensing:* is similar to the previous type of observation, but gives the participant more roles and control over the process. While many environmental and ecological observations follow data collection protocols that were designed by scientists, in participatory sensing the process is more distributed and emphasises the active involvement of the participants in setting what will be collected and analysed (e.g. WideNoise).

6. *Community/civic science:* also known as bottom-up science, is initiated and driven by a group of participants who identify a problem that is a concern for them and address it using scientific methods and tools. Within this type of activity, the problem, data collection and analysis are often carried out by community members or in collaboration with scientists or established laboratories (e.g. community-led air quality monitoring where they check if a local factory is polluting the environment nearby).

When examining the overlap between citizen science activities and geography, given that the activities of passive sensing, participatory sensing, environmental and ecological observations, and civic/community science inherently happen in a geographic location, they are inherently forms of VGI. When passive sensing focuses on health issues it might not use geographical information, and therefore it is depicted as potentially non-VGI. The case is more complex with volunteer computing and volunteer thinking, where projects do not necessarily deal with geographic information and can be about analysing neurons or looking at images of galaxies. Here, only when the issue is explicitly geographic – as in classifying images from a camera trap in the Serengeti – is the end result VGI.

STAKEHOLDERS AND ROLES IN VGI AND CITIZEN SCIENCE

Within the areas of VGI and citizen science, different actors take part in creating the systems that facilitate data collection and sharing. First, because of the complexity of setting up a system and running it efficiently, many VGI activities are considered to be a form of crowdsourcing and are run by companies who usually ensure that they preserve some or all intellectual property rights over the contributions. For example, although the images on Flickr are owned by the people who upload them to the service, Yahoo! holds some rights to the system, and can set what can be done and how they are accessed and queried with the images. As such there are few VGI projects established without explicit profit motives, and those that exist are run by volunteer-based organisations. OpenStreetMap, for example, is run by the OpenStreetMap Foundation.

The imbalance of power and access to information becomes clear as here each contributor is treated as an individual, and their contribution is minuscule compared to the totality of the information that the system owner has amassed through the effort of the participants. The system that facilitates crowdsourcing has costs associated with that provision, and for most participants, the contribution is comparable to the benefits that they extract from the system. For example, when a participant passively submits location information from a satellite navigation device, and the service provider delivers warnings of traffic delays on the basis of these reports, there is a balance between the minimal cost of sending the information and the service that the driver receives. Here, it would be correct to question whether or not the service provider should release all the information voluntarily provided by drivers at their own expense. And what of participants who invest significant amounts of time contributing to the project? Is it also appropriate to ask about the obligation of the service provider to such a contributor? The techno-libertarian answer would be that the volunteer does that of their own free will and the service provider does not have any responsibility to them. These are but a few moral dilemmas.

Citizen science is a more interesting and varied scenario in terms of the organisations that coordinate citizen science activities (see Haklay, 2015). To date, there has been little commercial interest in citizen science activities, and science is perceived as an activity that aims to improve the sum of human knowledge, even if there are specific individuals who benefit from this activity more than others. Historically, citizen science is an area that has involved many charities and non-governmental organisations (NGOs), and one of the most celebrated examples of citizen science – the Christmas Bird Count (see Goodchild, 2007) – is run by the National Audubon Society, a US not-for-profit dedicated to conservation. The role of volunteers in ecological observations and studies has increased in the past decade, as the scientific investment in ecology and long-running surveys by scientific institutions have decreased. Museums and other public engagement with science organisations are also active in citizen science, and it is part of their mission to educate and engage people with science. Researchers in universities and similar institutions engage with citizen science partly because it offers access to new resources that they would not be able to access otherwise without public support – for example, computing resources in the search for extra-terrestrial intelligence in the SETI@Home project. In some cases, such as large-scale classification tasks, there might be some similarities with the practice of crowdsourcing. Another issue within citizen science is access to the datasets that result from the efforts of volunteers. Here, there is a growing practice to share the results back with volunteers or the wider scientific community, to ensure that volunteers are credited in publications, as seen in efforts such as the Global Biodiversity Information Facility (GBIF[8]), which was set up to streamline the sharing of open and free biodiversity data.

Community science, as a form of citizen science, is of special interest, as it provides an example where new forms of factual spatial data are created and used to progress community goals. The Public Laboratory of Open Technology and Science[9] illustrates this. Public Lab is a community of environmental activists and technology experts that promotes the use of low-cost adapted ('hacked') technology to monitor environmental issues (Dosemagen et al., 2011). One of their early efforts was the creation of an aerial imagery apparatus using a kite or balloon to carry a cheap digital camera to take a large set of images over a relatively small area. The images that the camera captured are then sorted and stitched together to create a continuous image over the area where the balloon or kite was flown. This large-scale imagery provides visible evidence that is then annotated with additional information to highlight community issues – for example, to provide evidence on how many participate in a demonstration, or the impact of a new road on a Palestinian village in Jerusalem.[10] In public lab work, affordable technology is combined with the community's effort to provide instructions and guidance, which in turn supports the efforts to inform about a situation of local concern. In such situations citizen science is a tool of empowerment in the political sense as it provides both 'hard evidence' that emerges from scientific instruments or sensing devices, and methodology which supports a specific narrative that is of importance to the people who put it forward. It becomes an accepted form of evidence-based decision-making.

DEVELOPMENT AND FUTURE DIRECTIONS

Even though VGI and citizen science have much longer histories, most of the attention from policy-makers, researchers and businesses has only been expressed in the past decade. Frequently, questions arise about the quality of the resulting information as well as the motivation of the participants (see Sieber and Haklay, 2015). As more evidence emerges to confirm the quality of the data and that participants' motivations is recognised as not being a mass recruitment exercise, attention can turn towards the compilation of longitudinal data collection. In some VGI activities the collected information is 'hyper-local' – making it only relevant to a small area in both space and time – for example, information about a traffic jam and its implication on navigational decisions. Yet, even this localised information has relevance at a wider scale. In most VGI datasets, and especially in the area of citizen science, there is a need to understand how the information changes over time. Thus, the activities in these fields have the duality of describing a snapshot of the world (capturing an observation at a specific time and place and recording it), yet because of the continuous sharing of the information, the dataset as a whole is always dynamic and in a state of change (see Perkins, 2014, for detailed analysis of the (im)mutability of OpenStreetMap).

The process in which the information is produced, controlled and shared demonstrates differences in the power relationship and in financial benefits. Concurrently, the ability to maintain the repository of information over time should receive more attention. For example, OpenStreetMap servers require regular operating system updates and effort as well as resources to deal with hardware failures. Sustainability requires an organisation, institution or a company to take responsibility. As a result, the control of the system (understood here in the wider sense and not just the hardware/software part of it) foregrounds issues of power, regulation and resources into these seemingly distributed, non-hierarchical activities. In addition, the process of data quality assurance requires oversight and moderation of more experienced and knowledgeable participants who check the information provided by novices. Over time, power relationships reveal themselves in the case of both VGI and citizen science.

As VGI and citizen science activities progress, the questions persist regarding data quality, the longevity of engagement, incentives and motivation of volunteers, as well as the nature of the types of participants. Some research examining differential power has begun (Sieber and Haklay, 2015); there is nonetheless plenty of scope to critically study VGI and citizen science. For example, there are different levels of inclusiveness in terms of who is involved in data collection and which areas are being monitored. There is also merit to further investigating organisational practices and cultural influence with regards to the recruitment and ability to retain the participants over time. Aspects of gender inequality are being discussed (Cooper and Smith, 2010; Stephens, 2013), while ethnic, socio-economic and age have received less attention. There is also scope to understand how wider politics and economic incentives lead to outcomes; for example, which thematic areas receive attention and funding and, more specifically, how is it that the production of base maps is perceived as a valuable commercial activity, while the recording of biodiversity is not? Understanding VGI and citizen science as a socio-technical system, and giving due attention to social aspects, might provide better insight about the nature of the spatial data being produced through these activities and what these tell us about the state of the world.

NOTES

1. http://www.opensignal.com/
2. http://www.yelp.com/
3. http://www.wikipedia.org/
4. http://www.streetbump.org/
5. http://www.openstreetmap.org/
6. http://qcn.stanford.edu/
7. http://geotagx.org/
8. http://www.gbif.org/
9. http://publiclab.org/
10. http://publiclab.org/wiki/jerusalem

REFERENCES

Becker, M., Caminiti, S., Fiorella, D., Francis, L., Gravino, P., Haklay, M., Hotho, A., Loreto, V., Mueller, J., Ricchiuti, F., Servedio, V.D., Sîrbu, A. and Tria F. (2013) 'Awareness and learning in participatory noise sensing', *PLoS One*, 8 (12): e81638.

Brown, J. (2001) 'Three case studies', in C. Werry and M. Mowbray (eds), *Online Communities: Commerce, Community Action, and the Virtual University*. Upper Saddle River, NJ: Prentice Hall.

Calabrese Barton, A.M. (2012) 'Citizen(s') science: a response to "The Future of Citizen Science"', *Democracy and Education*, 20 (2): article 12. Available at: http://democracyeducationjournal.org/home/vol20/iss2/12 (accessed 26 July 2016).

Cooper, C.B. (2012) 'Links and distinctions among citizenship, science, and citizen science: a response to "The Future of Citizen Science"'. *Democracy and Education*, 20 (2): article 13. Available at: http://democracyeducationjournal.org/home/vol20/iss2/13 (accessed 26 July 2016).

Cooper, C.B. and Smith, J.A. (2010) 'Gender patterns in bird-related recreation in the USA and UK', *Ecology and Society*, 15 (4): 4.

Craglia, M., Ostermann, F. and Spinsanti, L. (2012) 'Digital Earth from vision to practice: making sense of citizen-generated content', *International Journal of Digital Earth*, 5 (5): 398–416.

Dosemagen, S., Warren, J. and Wylie, S. (2011) 'Grassroots mapping: creating a participatory map-making process centered on discourse', *Journal of Aesthetics and Protest*, 8. Available at: http://www.joaap.org/issue8/GrassrootsMapping.htm (accessed 26 July 2016).

Goodchild, M.F. (2007) 'Citizens as sensors: the world of volunteered geography', *GeoJournal*, 69 (4): 211–21.

Haklay, M. (2013) 'Citizen science and Volunteered Geographic Information – overview and typology of participation', in D.Z. Sui, S. Elwood and M.F. Goodchild (eds), *Crowdsourcing Geographic Knowledge*. Berlin: Springer. pp. 105–22.

Haklay, M. (2015) *Citizen Science and Policy: A European Perspective*. Washington, DC: Woodrow Wilson Center for International Scholars.

Haklay, M., Singleton, A. and Parker, C. (2008) 'Web mapping 2.0: the neogeography of the geoweb', *Geography Compass*, 3: 2011–39.

Howe, J. (2006) 'The rise of crowdsourcing', *Wired*, 14 (6).

Laurel, B. (2001) *Utopian Entrepreneur*. Cambridge, MA: MIT Press.

Mueller, M.P., Tippins, D. and Bryan, L.A. (2012) 'The future of citizen science', *Democracy and Education*, 20 (1): article 2. Available at: http://democracyeducationjournal.org/home/vol20/iss1/2 (accessed 26 July 2016).

Oxford English Dictionary (OED) (2014) 'Citizen science' entry. Oxford: Oxford University Press.

Perkins, C. (2014) 'Plotting practices and politics: (im)mutable narratives in Open StreetMap', *Transactions of the Institute of British Geographers*, 39 (2): 304–17.

Shirky, C. (2011) *Cognitive Surplus: Creativity and Generosity in a Connected Age*. New York: Penguin.

Sieber, R.E. and Haklay, M. (2015) 'The epistemology(s) of volunteered geographic information: a critique', *Geo: Geography and Environment*, 2 (2): 122–36.

Silverman, J. (2014) 'The crowdsourcing scam: why do you deceive yourself?', *The Baffler*, 26. Available at: http://www.thebaffler.com/salvos/crowdsourcing-scam (accessed 26 July 2016).

Stephens, M. (2013) 'Gender and the geoweb: divisions in the production of user-generated cartographic information', *GeoJournal*, 78 (6): 981–96.

13

GEO-SEMANTIC WEB

PETER L. PULSIFER AND GLENN BRAUEN

INTRODUCTION

The internet, and the World Wide Web (the web) in particular, have revolutionised how humans share information. Massive volumes of data are now shared through the internet, with an estimated throughput of 60 exabytes per month (one exabyte = 1 billion gigabytes) and over 14 billion devices (as of 2014) (CISCO Systems, 2015). While information exchange is pervasive, interoperability between networked systems is not universal. In a very general sense, interoperability refers to the ability to readily share information and/or operations for a particular purpose across information products or systems, present or future, without unintended restrictions. Interoperability presents the possibility of linking and integrating data and information held in two, tens or many thousands of systems with the potential for increased efficiency (e.g. less duplication of data), and enabling innovative and valuable new applications including new kinds of science (Hey et al., 2009). Full interoperability goes beyond simply connecting systems to enable the effective interchange and use of data; it includes understanding the meaning of the data. The internet and the web very effectively support the technical aspects of exchanging data and information; however, they fall short in providing an understanding of the meaning of data or information: semantics. This chapter focuses on broad efforts to develop a 'Semantic Web' – and, more specifically, a Geo-Semantic Web – to address the challenges of semantic interoperability for spatial data and media.

INTEROPERABILITY AND THE (GEO)SEMANTIC WEB

In decades past, interoperability between systems was very limited. It was difficult to share digital information between computers with different operating systems (e.g. Windows, Macintosh), or characters from different languages (e.g. English, Chinese). These challenges have mostly been addressed through standardisation of file structure and international character-sets. However, many other types of interoperability remain, including but not limited to structural, semantic, legal and cultural, with semantic interoperability remaining as one of the major challenges. Semantic interoperability can be defined as the ability to effectively exchange meaning between information systems. Semantics define concepts and terms and can also store information about the relationships between concepts. While semantic interoperability has been achieved in some systems that serve very well-defined purposes and information needs (e.g. banking, travel information), challenges remain in broader, more general contexts (e.g. science). Semantic differences are difficult to reconcile in a machine environment. People can be quite good at dealing with the multiple meanings of language in different contexts. Computers, on the other hand, function best when words are always used to mean the same thing and the relationships between words are clear.

One approach to help computers deal with semantic interoperability is the (Geo) Semantic Web (Berners-Lee et al., 2001; Fonseca, 2008). The Semantic Web has been evolving for decades, with standards being developed in the late 1990s (Gruber, 1993; Powers, 2003). Consequently, the Semantic Web is maturing and now underpins a number of scientific cyberinfrastructure initiatives (e.g. Fox et al., 2009; Ribes and Bowker, 2009). Moreover, software development methods using the Semantic Web are becoming well-documented and common outside of academia (Segaran et al., 2009). For many years, Geo-Semantic Web research has also been prominent (Bishr, 1998; Agarwal, 2005) and, recently, media studies have adopted semantics as a significant area of research and development (Stamou and van Ossenbruggen, 2006).

THE GEO-SEMANTIC WEB IN PRACTICE

The Geo-Semantic Web uses models, each containing concepts and associated terms to link similar data, to mediate differences in meaning between knowledge systems and even potentially generate new knowledge through reasoning. These knowledge models are called 'formal ontologies' or simply 'ontologies'. An ontology that includes information about ocean shipping routes and their impact on Inuit harvesting practices, for example, might contain statements such as:

- A shipping route is the average of multiple ship tracks.

- A shipping route may disrupt Inuit travel patterns and harvesting activities.

- A shipping route may prevent access to harvesting areas.

- A shipping route passes through Anaktalak Bay.

Ontologies, as discussed here, describe, expose, share and connect pieces of data through the use of unique persistent identifiers and standardised protocols. In practice, the encoding of ontologies uses standard structures and languages such as the

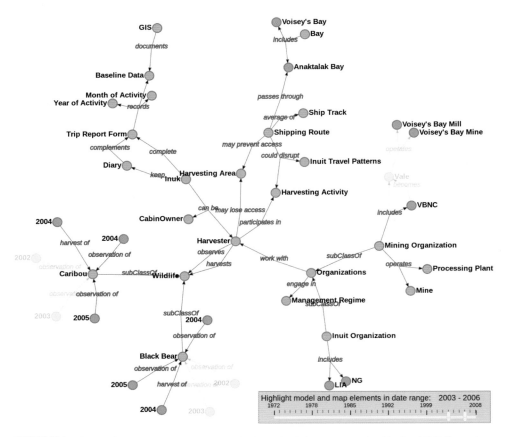

FIGURE 13.1   Simplified Nunatsiavut Knowledge Model composed of linked RDF triples

Resource Description Framework (RDF; W3C, 2014) and the Web Ontology Language (OWL; W3C, 2004). These mechanisms represent knowledge as a set of assertions, each called a triple. A triple comprises two conceptual nodes (subject, object) joined by a connecting edge (predicate). The two nodes plus an edge make up a triple; for example, {shipping route; passes through; Anaktalak Bay} (Figure 13.1, near top right).

Data can be stored using different structures: a single table as with a spreadsheet, or a set of related tables commonly used in a spatial–relational database, or even unstructured data within a document. Combining data stored using different structural models requires manual, semi-automatic or automatic (if possible) restructuring before use. The triple model supports structural interoperability by decomposing tables of data into individual units of data (i.e. triples) that can each be linked to or combined with other units of data to form equivalent knowledge structures to what we think of as tables, databases or other structures. One or more ontologies can be used to provide structure to data units. Within an ontology, subject, object and predicate may be designated with 'local' names or they may re-use appropriate identifiers selected from known, published vocabularies. Vocabularies are agreed collections of terms, usually related to a specific topic, and most generally defined and published using persistent universal resource identifiers (URIs) (e.g. Table 13.1). When used in modelling, these vocabulary terms promote the re-use of ontologies by allowing for the creation of indexed, searchable access to the triples that make up an ontology (Hyland et al., 2014).

Used with ontologies, semantic tools can be applied to mediate across different data structures, assisting us with tasks such as searching for relevant elements within large collections of triples. At the level of semantics, if an ontology is sufficiently expressive, Geo-Semantic Web tools can also deal with semantic interoperability problems, such as terms that have been defined differently by communities wanting to share data. For example, if a term has many different synonyms an ontology can use a 'same as' relationship to translate data that contain different but synonymous terms. A detailed explanation of how Geo-Semantic Web tools work is beyond the scope of this chapter; however, standardised syntax elements, the structural simplicity of triples, and the structural and semantic information contained in an ontology can all be combined to support interoperability across systems.

STANDARDS AND TOOLS

Many standards and tools are now available to construct ontologies, and serve and manipulate semantic data (e.g. RDF, OWL; Apache Jena, 2015; Parliament, 2015). Triples are typically stored in a semantic database known as a triple-store and published over the internet (cf. Linked Data Cloud, 2015). Triple-stores can be queried using a formal language such as SPARQL (SPARQL Protocol and RDF Query Language). Major industry-academic-standards initiatives are fully engaged in developing geospatial web standards and technologies to deal with geographic linked open data (LOD: OGC, 2015; W3C, 2015). Specifically, the Open Geospatial Consortium (OGC) leads a prominent initiative to develop GeoSPARQL, a combination of defined geospatial data vocabulary represented in RDF and an extension to the general Semantic Web SPARQL query language. GeoSPARQL supports the specific requirements of processing geospatial data (e.g. supporting queries based on spatial measurements, such as distance, and relations such as 'is contained by'). Using GeoSPARQL as a query language, users can query triples to filter or reason for particular purposes, for example: select all features that have intersecting geometries of a specific type (e.g. shipping routes that intersect particular bays).

An ontology may include knowledge of hierarchical relationships (this subject is a parent of that object), semantic relationships (this subject term means or does not mean the same as that object term) and simple data constraints (the value of this object cannot exceed 100). LOD data can be used without an associated ontology, however an ontology expresses data context and meaning. These standards and tools provide a new platform for application development focusing on a range of topics and can include geographic concepts and ideas, and media concepts and relations.

THE EVOLVING GEO-SEMANTIC WEB

Possibilities aside, the Geo-Semantic Web is not mainstream. Popular GIS software packages, such as ArcGIS and Quantum GIS, do not yet have well-developed interfaces to Semantic Web models and data, although limited support is emerging. Thus, by extension, semantic spatial media have not yet become mainstream, and this situation suggests that the slow rate of adoption is due in part to the complexity of the Semantic Web model and the stack of required standards and technologies.

The Semantic Web vision is built on a relatively complex architectural model. This model starts with the use of a standardised syntax element such as Unicode, and includes the use of Extensible Markup Language (XML) layers as a way of encoding data, and then adds RDF, OWL and optionally other standards related to security and applying rules and logic to ontologies. The resulting model requires that practitioners be familiar with many standards and tools, and use particular encodings (i.e. XML). While implementation using this model may be achievable for IT professionals in well-resourced organisations, participation by other practitioners such as students, individual or small groups of researchers, artists, developers at small companies and other can be limited. In our work developing applications with Arctic communities, it has been observed and experienced that the teams and budgets involved are typically small, limiting the model's adoption. Furthermore, applications are often developed under discrete project funding rather than an ongoing, long-term and sustainable resource model which results in developers, researchers and managers avoiding overly complex development models.

In response to the complexity of the original Geo-Semantic Web model, a second generation is emerging. The second generation comprises a new set of technologies, methods and architectural patterns that enable information systems to be built on the original Geo-Semantic Web model, but use modified implementation practices. These emerging development approaches correspond to concurrent developments in browser-based computing, use of microformats for metadata tagging in web pages, the ubiquity of mobile devices, the dominance of social media, and developments in distributed and cloud computing. The new developments enable practitioners and developers who do not have advanced degrees in information sciences, or extensive experience with first-generation Geo-Semantic Web technologies, to participate in the Geo-Semantic Web movement.

INITIAL EXPERIENCES IN APPLYING CURRENT GEOSPATIAL WEB DEVELOPMENTS

REPRESENTING INDIGENOUS KNOWLEDGE

So far we have provided a summary discussion of the opportunities and challenges of using Geo-Semantic Web methods and technologies to achieve semantically interoperable spatial media. We will now illustrate the argument with reference to the documentation of Arctic Region Indigenous Knowledge. Indigenous Knowledge can be defined as:

knowledge and know-how accumulated across generations, and renewed by each new generation, which guide human societies in their innumerable interactions with their surrounding environment. (Nakashima et al., 2012).

Indigenous Knowledge is holistic (does not separate nature/culture/spirituality), complex, dynamic, geographically and culturally situated, and spatiotemporal, and it is challenging to represent all of these elements through documentation in a nuanced way (Johnson et al., 2015). Consequently, representing Indigenous Knowledge in digital form often includes the use of multimedia such as georeferenced digital photos, audio, video and documents (Brauen et al., 2011; Pulsifer et al., 2011, 2012, 2014; Hayes et al., 2014). In the remainder of the chapter we provide preliminary insights resulting from the ongoing development of a spatial media application that uses second-generation Geo-Semantic Web standards and tools.

LINKING CONCEPTS, GEOGRAPHIC INFORMATION AND MEDIA

To support modelling projects which combine Indigenous Knowledge, geographic information collected over a number of years, and multimedia items that provide context and additional information about the models, the authors in collaboration with other researchers are developing a new, lightweight set of tools to create and visualise ontologies. These models include and link concepts, particular instances of concepts, relationships among these, mapped geographical instances, and multimedia elements. This approach is distinguished in comparison to other applications that combine concept mapping, mapping, multimedia and time (e.g. Hayes et al., 2014) by striving to establish a model that contains 'global' concepts and object naming where available, rather than project-specific, local names. A primary mechanism for doing this is the use of geospatial web ontologies shared with a broader community using linked open data (LOD) as a mechanism for standardising data exchange.

Figure 13.1 is a simplified model that describes aspects of Inuit harvesting along with related institutional and policy frameworks and potential impacts arising from resource activities such as mining in Nunatsiavut (northern Labrador). Along with potential shipping-route impacts, black bears and caribou are presented as subclasses of wildlife with annual harvest and observation records from Inuit harvesters (lower left). Maps showing each of the annual harvest and observation datasets can be displayed by selecting the subject node of an 'observation of' or 'harvest of' triple, labelled with the year (e.g. 2004), and selecting an option to view the map of that data. This requires that a user interacting with the model be able to initiate a spatio-temporal selection of some mappable data (e.g. 'Show me the caribou observations for 2004'). Similarly, time can be factored into the visualisation of the model directly through the use of temporal brushing by using a time slider (Monmonier, 1990). In Figure 13.1, the time slider has been adjusted so that wildlife harvest and observation data prior to 2004 are de-emphasised (lower left) and only the pre-2007 ownership of the Voisey's Bay mine by the Voisey's Bay Nickel Company (VBNC) is highlighted (right).

Encoding data within ontologies usually draws on existing schemas and vocabularies, possibly extending those through the creation of additional definitions and relations within concepts being modelled. Re-use of known ontologies and vocabularies, when appropriate, can be an efficient way both to link new models to existing datasets and to avoid the reinvention of schemas and terminologies in support of interoperability. The integration of previously defined ontologies and vocabularies is an active research area resulting, in part, from the large number of existing ontologies and the complexity of coming to full agreement on the definition of concepts and terms contained in ontologies that are part of the Linked Open Data cloud (Jain et al., 2010; Parundekar et al., 2012). For example, the Marine Metadata Initiatives lists many ontologies, many of which cover similar domains (e.g. sensors, biological terms) but are developed by different groups (see https://marinemeta data.org/conventions/ontologies-thesauri).

Although some ontologies are intended to make commonly required definitions such as time (Hobbs and Pan, 2006), easy to re-use in domain-specific models, many existing ontologies have been designed for particular uses, contain only the concepts and relations needed for those, and are therefore difficult to re-use, hampering interoperability. In general, finding appropriate ontologies for re-use is neither obvious nor trivial (Noy, 2004). Re-using vocabularies alone, although inadequate as a long-term approach because it will still allow and result in some proliferation of similar ontologies, at least re-uses existing object definitions rather than creating additional terms. To this end, we have identified vocabularies we could re-use to link or bind concepts, geographical information and multimedia together, requiring us only to create project-specific vocabularies as needed to fill gaps. This is a pragmatic approach to the imperfect but advancing state of geo-semantic modelling. Table 13.1 provides a summary by subject heading of possible ontologies that could provide stable (URI-based) vocabularies which clarify instances and, if only indirectly, the concepts in the ontology fragments in Figure 13.1. In addition to concepts from the existing prototype, stable instance identifiers for documentation or publications that flesh out policy regimes and/or research related to policy may be useful and are included in the table.

Framework support to integrate references to media objects, such as books, into geo-semantic ontologies would be highly desirable. In addition to text, Indigenous Knowledge is commonly recorded in unstructured sources of media, such as video and audio recordings. Ontologies and vocabularies for describing published works in a variety of media types are now widely adopted by library catalogues (OCLC, 2014; Papadakis et al., 2015), broadcast information systems (Kobilarov et al., 2009), and social media firms (Facebook, 2015) among others. Dublin Core (Rühle et al., 2011), for example, provides semantics for describing digital publications generally and, although originally intended for texts, has been broadened 'to encompass, in principle, any object that can be identified, whether electronic, real-world, or conceptual'. Dublin Core elements describe an object in terms of content (e.g. title, description, audience, and geographic coverage), intellectual property (e.g. creator, publisher) and particulars such as language, date and format of an instance of a creative work (Hillman, 2005).

Other aspects of digital media can be described through the use of additional ontologies designed to work with and extend Dublin Core. For example, Lee et al. (2012) defined a media ontology with usage guidelines for a variety of media and metadata

TABLE 13.1 Existing linkable vocabularies that could be used in ontology revisions

Subject	Vocabulary	Example(s)
Organisations/ Stakeholders	LCNAF (Library of Congress, 2011a)	Labrador Inuit Association, VBNC
Publications	Worldcat.org (OCLC, 2014) VIAF (2015)	*Our Footprints are Everywhere* (Brice-Bennett, 1977)
Journal articles	CrossRef.org (CrossRef, 2011)	Digital object identifier based URI generator for consortium of journal publishers
Species	GeoSpecies (DeVries, 2013) LCSH (Library of Congress, 2011b)	Caribou, black bear
Time	Time (Hobbs and Pan, 2006)	Dates, date ranges
Geospatial	GeoNames (http://www.geonames.org/ ontology/documentation.html) GeoSPARQL (http://www.opengeospatial. org/standards/geosparql)	Place names; geographic features (concept + geometry)

formats including the Exif metadata format used to describe digital photos and the more complex encoding of the MPEG-7 multimedia metadata specifications (see Chaudhury et al., 2015). Franz et al. (2011) proposed the Core Ontology for Multimedia (COMM), based on the capabilities of the MPEG-7 standard, to support new media annotation requirements such as identifying geometric portions of a video frame (e.g. the bounding box around a person's face) within a sequence from a longer video. For image, audio and video, Schallauer et al. (2011) provide a recent review of metadata standards.

CHALLENGES TO THE FULL ADOPTION OF THE GEO-SEMANTIC WEB MODEL IN THIS CASE

When considering the use of ontologies in representing Indigenous Knowledge through spatial media, we have identified two key impediments to the full adoption of a geospatial web model. First, while many ontologies exist and new ontologies are appearing regularly, many concepts and relationships are application specific. The Semantic Web is too new to have published URIs for all content that might be included in an application. Much work remains in encouraging information communities to publish and model in this way, and there are still questions about the appropriateness of doing this in some cases. Second, for the Semantic Web model to be useful, relevant instances of data must be published using a Linked Open Data model and they must be found by people creating new models. While adoption is increasing, much of the content on the web still does not follow these practices. The movement towards Linked Open Data needs to mature before full interoperability potential can be realised. In addition, tools are required to assist model designers in finding and linking existing vocabulary references where appropriate and in helping them to create and publish linkable vocabularies as needed.

Addressing these is a matter of promoting a paradigm shift in how data are modelled and stored. This is happening, but slowly. However, the increasing prominence of the Geo-Semantic Web in research and industry indicates that this process is accelerating. The emergence of effective tools for working with ontologies and vocabularies is helping this process as the GSW moves towards the mainstream (e.g. JSON-LD: http://json-ld.org/).

CONCLUSION

The Geo-Semantic Web is a flexible, powerful and interoperable framework for developing spatial media applications. However, it requires the formalisation of knowledge through the establishment of triples and ontologies. Critical reflection reveals that formalism is not appropriate in all contexts. Ontology development is often driven by a set of ideal design criteria that establish a 'good' ontology (Gruber, 1993: 17). The characteristics of a good ontology include objective definitions independent of social context; definitions stated in logical axioms; complete definitions; no logical inconsistency; no dependence on the symbols used to encode it (i.e. language independent); and the minimum number of claims required to support intended knowledge sharing (Gruber, 1993). There are models of knowledge and cognition that do not conform to the rules established for formal ontology. For example, through the lens of cognitive linguistics, the formal ontology rules presented here nonetheless suggest an objectivist and essentialist view (Lakoff, 1987). This does not allow for the nuances and potential inconsistencies inherent in culture and cognition (Levinson, 2003), the nature of cognitive models (Weiskopf, 2009), and the nature and use of metaphor in language (Lakoff and Johnson, 2008). As Wilks (2014) argues, different philosophical approaches to meaning lead to divergent conclusions about how semantics will develop and grow within the Semantic Web, with possibilities including controlled formal ontologies (similar to the original visions of the Semantic Web), and more fragmented, informal and somewhat democratic proliferations of ontologies developing more similarly to languages in which meanings are intertwined with usage.

Thus, complete agreement does not exist between the semantic model and social and cognitive frameworks. Here we suggest that appropriate components of the semantic model can be adopted to improve our ability to represent socially and cognitively complex knowledge (e.g. Indigenous Knowledge) across domains. These components are:

- use of distributed linked data including URIs;
- extensive support for modelling rich relationships;
- extendibility;

- some level of language independence;
- use of ontology language to partially model knowledge domain.

Components of the Semantic Web that were deemed less or not appropriate in this context are:

- definitions that are independent of social context;
- definitions stated solely in logical axioms;
- assumptions that a single, authoritative ontology can be established even within a specific information community;
- use of formal logic for all inference.

A major limitation of the first generation of the Semantic Web technologies was the reliance on representation and reasoning that is optimal for highly structured or textual data. Criticisms focus on the difficulty of formal classification in environments that are informal, highly distributed, lacking central coordination and authorities, and including heterogeneous and complex information (Millerand and Bowker, 2009). These concerns must be considered when using multimedia data and the knowledge they represent. Research and development is required to establish the limits and appropriateness of using the Geo-Semantic Web with spatial media.

We conclude that the Geo-Semantic Web has great potential but is still evolving. The flexible structure of triples provides a powerful method for supporting interoperability. The potential for linking a broad range of different data types and sources using data in this form is enormous. The use of ontologies and their ability to create knowledge models that link data nodes (again, using triples) establishes a very powerful model for identifying concepts and relationships. Conceivably, any number of ontologies can be applied to the same data, resulting in a pluralistic modelling environment. However, Geo-Semantic Web and semantic spatial media are still not mainstream. The complexity of the Geo-Semantic Web approach and underlying model and tools is an impediment to adoption. Designed in the late 20th century, the standards, formats and methods involved make for a steep and time-consuming learning curve that requires very specialised knowledge. The second-generation Geo-Semantic Web – based on small, emergent components derived from the larger, more complex vision – promises a more accessible set of tools.

In working to use the Geo-Semantic Web to represent Indigenous Knowledge (commonly using spatial and media data), we have established that there are some resources available in terms of ontologies, but these are at the early stages of maturity. There are few established ontologies for multimedia in particular. Similarly, ontologies for Indigenous Knowledge, and particularly that of the Arctic, are not yet readily available.

There is an increasingly large array of Geo-Semantic Web tools available; however, many are still difficult to use or are not readily useful in web application development, a primary platform for Indigenous Knowledge presentation. This is changing with the advent of new data encodings (e.g. JSON-LD) and the resulting ability to use Geo-Semantic Web data and methods in a web development environment.

REFERENCES

Agarwal, P. (2005) 'Ontological considerations in GIScience', *International Journal of Geographical Information Science*, 19 (5): 501–36.

Apache Jena (2015) 'Apache Jena: a free and open source Java framework for building Semantic Web and Linked Data applications'. Available at: https://jena.apache.org (accessed 26 July 2016).

Berners-Lee, T., Hendler, J. and Lassila, O. (2001) 'The Semantic Web', *Scientific American*, 284 (5): 34–44.

Bishr, Y.A. (1998) 'Overcoming the semantic and other barriers to GIS interoperability', *International Journal of Geographical Information Systems*, 12 (4): 299–314.

Brauen, G., Pyne, S., Hayes, A., Fiset, J.P. and Taylor, D.R.F. (2011) 'Encouraging transdisciplinary participation using an open source cybercartographic toolkit: the Atlas of the Lake Huron Treaty Relationship Process', *Geomatica*, 65 (1): 27–45.

Brice-Bennett, C. (1977) *Our Footprints are Everywhere: Inuit Land Use and Occupancy in Labrador*. Nain, Newfoundland: Labrador Inuit Association.

Chaudhury, S., Mallik, A. and Ghosh, H. (2015) *Multimedia Ontology: Representation and Applications*. London: Chapman and Hall/CRC.

CISCO Systems (2015) 'VNI forecast highlights', *Cloud and Mobile Network Traffic Forecast*. Available at: http://www.cisco.com/web/solutions/sp/vni/vni_forecast_highlights/index.html (accessed 26 July 2016).

CrossRef (2011) 'CrossRef and International DOI Foundation collaborate on linked-data-friendly DOIs', CrossRef Blog. Available at: http://www.crossref.org/crweblog/2011/04/crossref_and_international_doi.html (accessed 26 July 2016).

DeVries, P.J. (2013) 'The Datahub: GeoSpecies knowledge base'. Available at: http://datahub.io/dataset/geospecies (accessed 26 July 2016).

Facebook (2015) 'The Graph API'. Available at: https://developers.facebook.com/docs/graph-api (accessed 26 July 2016).

Fonseca, F. (2008) 'The geospatial semantic web', in J.P. Wilson and A.S. Fotheringham (eds), *The Handbook of Geographic Information Science*. Malden, MA: Blackwell Publishing. pp. 388–91.

Fox, P., McGuinness, D.L., Cinquini, L., West, P., Garcia, J., Benedict, J.L. and Middleton, D. (2009) 'Ontology-supported scientific data frameworks: the virtual solar-terrestrial observatory experience', *Computers and Geosciences*, 35 (4): 724–38.

Franz, T., Troncy, R. and Vacura, M. (2011) 'The core ontology for multimedia', in R. Troncy, B. Huet and S. Schenk (eds), *Multimedia Semantics*. Chichester: John Wiley and Sons. pp. 145–61.

Gruber, T.R. (1993) 'A translation approach to portable ontology specifications', *Knowledge Acquisition*, 5 (2): 199–220.

Hayes, A., Pulsifer, P.L. and Fiset, J.P. (2014) 'The Nunaliit cybercartographic atlas framework', in D.R.F. Fraser and T.P. Lauriault (eds), *Developments in the Theory and Practice of Cybercartography*. Amsterdam: Elsevier. pp. 129–40.

Hey, A.J.G., Tansley, S. and Tolle, K.M. (2009) *The Fourth Paradigm: Data-Intensive Scientific Discovery*. Redmond, WA: Microsoft Research.

Hillman, D. (2005) 'Using Dublin Core'. Available at: http://dublincore.org/documents/usageguide/ (accessed 26 July 2016).

Hobbs, J.R. and Pan, F. (2006) 'Time ontology in OWL', W3C working draft. Available at: https://www.w3.org/TR/owl-time/ (accessed 26 July 2016).

Hyland, B., Atemezing, G. and Villazon-Terrazas, B. (2014) 'Best practices for publishing linked data', W3C Working Group Note, January: 1–22. Available at: http://www.w3.org/TR/ld-bp/ (accessed 26 July 2016).

Jain, P., Hitzler, P., Sheth, A.P., Verma, K. and Yeh, P.Z. (2010) 'Ontology alignment for linked open data', *Semantic Web–ISWC 2010: 9th International Semantic Web Conference, ISWC 2010*, Shanghai, China, November 7–11, 2010, Revised Selected Papers, Part I. Springer: Berlin. pp. 402–17.

Johnson, N., Alessa, L., Behe, C., Danielsen, F., Gearheard, S., Gofman-Wallingford, S., Kliskey, A. et al. (2015) 'The contributions of community-based monitoring and traditional knowledge to Arctic observing networks: reflections on the state of the field', *Arctic*, 68 (1): 1–13.

Kobilarov, G., Scott, T., Raimond, Y., Oliver, S., Sizemore, C., Smethurst, M., Bizer, C. and Lee, R. (2009) 'Media meets semantic web – how the BBC uses DBpedia and linked data to make connections', in L. Aroyo, P. Traverso, F. Ciravegna, P. Cimiano, T. Heath, E. Hyvönen, R. Mizoguchi, E. Oren, M. Sabou and E. Simperl (eds), *The Semantic Web: Research and Applications*. Springer: Berlin. pp. 723–37.

Lakoff, G. (1987) *Women, Fire, and Dangerous Things: What Categories Reveal About the Mind*. Chicago, IL: University of Chicago Press.

Lakoff, G. and Johnson, M. (2008) *Metaphors We Live By*. Chicago, IL: University of Chicago Press.

Lee, W., Bailer, W., Bürger, T., Champin, P.-A., Evain, J.-P., Malaisé, V. and Strassner, J. (2012) 'Ontology for media resources 1.0', W3C Recommendation, February. Available at: http://www.w3.org/TR/mediaont-10/ (accessed 26 July 2016).

Levinson, S.C. (2003) *Space in Language and Cognition: Explorations in Cognitive Diversity*. Cambridge: Cambridge University Press.

Library of Congress (2011a) 'Library of Congress names - LC linked data service'. Available at: http://id.loc.gov/authorities/names.html (accessed 26 July 2016).

Library of Congress (2011b) 'Library of Congress subject headings - LC linked data service'. Available at: http://id.loc.gov/authorities/subjects.html (accessed 26 July 2016).

Linked Data Cloud (2015) 'Linked data - connect distributed data across the web'. Available at: http://linkeddata.org/ (accessed 26 July 2016).

Millerand, F. and Bowker, G. (2009) 'Metadata standards: trajectories and enactment in the life of an ontology', in M. Lampland and L.S. Star (ed.), *Standards and their Stories*. Ithaca, NY: Cornell University Press. pp. 149–76.

Monmonier, M. (1990) 'Strategies for the visualization of time-series data', *Cartographica*, 27 (1): 30–45.

Nakashima, D.J., Galloway McLean, K., Thulstrup, H.D., Ramos Castillo, A. and Rubis, J.T. (2012) *Weathering Uncertainty: Traditional Knowledge for Climate Change Assessment and Adaptation*. Paris: Darwin.

Noy, N.F. (2004) 'Semantic integration: a survey of ontology-based approaches', *ACM Sigmod Record*, 33 (4): 65–70.

OCLC (2014) 'OCLC releases WorldCat works as linked data', OCLC. Available at: https://www.oclc.org/news/releases/2014/201414dublin.en.html (accessed 26 July 2016).

Open Geospatial Consortium (OGC) (2015) 'GeoSPARQL - a geographic query language for RDF data', OGC. Available at: http://www.opengeospatial.org/standards/geosparql (accessed 26 July 2016).

Papadakis, I., Kyprianos, K. and Stefanidakis, M. (2015) 'Linked data URIs and libraries: the story so far', *D-Lib Magazine*, 21 (5/6). Available at: http://www.dlib.org/dlib/may15/papadakis/05papadakis.html (accessed 26 July 2016).

Parliament [software] (2015) 'Parliament High-Performance Triple Store'. Available at: http://parliament.semwebcentral.org (accessed 26 July 2016).

Parundekar, R., Knoblock, C.A. and Ambite, J.L. (2012) 'Discovering concept coverings in ontologies of linked data sources', *The Semantic Web–ISWC 2012: 11th International Semantic Web Conference, Boston, MA, USA, November 11–15, Proceedings, Part I*. Berlin: Springer. pp. 427–43.

Powers, S. (2003) *Practical RDF*. Sebastapol, CA: O'Reilly.

Pulsifer, P.L., Laidler, G.J., Taylor, D.R.F. and Hayes, A. (2011) 'Towards an indigenist data management program: reflections on experiences developing an Atlas of Sea Ice Knowledge and Use', *Canadian Geographer*, 55 (1): 108–24.

Pulsifer, P., Gearheard, S., Huntington, H.P., Parsons, M.A., McNeave, C. and McCann, H.S. (2012) 'The role of data management in engaging communities in Arctic research: overview of the exchange for local observations and knowledge of the Arctic (ELOKA)', *Polar Geography*, 35 (3–4): 271–90.

Pulsifer, P.L., Huntington, H.P. and Pecl, G.T. (2014) 'Introduction: local and traditional knowledge and data management in the Arctic', *Polar Geography*, 37 (1): 1–4.

Ribes, D. and Bowker, G. (2009) 'Between meaning and machine: learning to represent the knowledge of communities', *Information and Organization*, 19 (4): 199–217.

Rühle, S., Baker, T. and Johnston, P. (2011) 'User guide - DCMI MediaWiki', Dublin Core Metadata Initiative. Available at: http://wiki.dublincore.org/index.php/User_Guide (accessed 26 July 2016).

Schallauer, P., Bailer, W., Troncy, R. and Kaiser, F. (2011) 'Multimedia metadata standards', in R. Troncy, B. Huet and S. Schenk (eds), *Multimedia Semantics*. Chichester: John Wiley and Sons. pp. 129–44.

Segaran, T., Evans, C. and Taylor, J. (2009) *Programming the Semantic Web*. Santa Clara, CA: O'Reilly Media, Inc.

Stamou, G. and van Ossenbruggen, J. (2006) 'Multimedia annotations on the semantic web', *IEEE MultiMedia*, 13 (1): 86–90.

VIAF (2015) 'Virtual International Authority File'. Available at: http://viaf.org/viaf/data/ (accessed 26 July 2016).

Weiskopf, D.A. (2009) 'The plurality of concepts', *Synthese*, 169 (1): 145–73.

Wilks, Y. (2014) 'Beyond the internet and web', in M. Graham and Dutton, W.H. (eds), *Society and the Internet*. Oxford: Oxford University Press. pp. 360–72.

W3C (2004) 'OWL web ontology language overview'. Available at: http://www.w3.org/TR/owl-features/ (accessed 26 July 2016).

W3C (2014) 'Resource Description Framework (RDF)'. Available at: http://www.w3.org/RDF/ (accessed 26 July 2016).

W3C (2015) 'Geospatial semantic web community group'. Available at: http://www.w3.org/community/geosemweb/wiki/Main_Page (accessed 26 July 2016).

14

SPATIAL DATA ANALYTICS

HARVEY J. MILLER

INTRODUCTION

Spatial media are both consumers and producers of vast quantities of spatial big data. A key way of making sense of such data is to use spatial data analytics, which is the data-driven process of generating better scientific, planning and management decisions from georeferenced data. 'Data-driven' means that measurements of spatial phenomena drive the nature of the analysis: the data are not only a way to calibrate, validate and test a spatial model but rather the force behind the analysis. Consequently, analysts design spatial data analytical techniques with data in mind (Miller and Goodchild, 2014). This is similar to the definition of geocomputation by Fotheringham (1998): the computer is not just a convenient mechanism but rather changes how the analysis occurs.

We often hear about the so-called three Vs of big data – volume, variety and velocity (Dumbill, 2012) – with some adding veracity and value and others adding exhaustivity, resolution, indexicality, relationality, extensionality and scalability (Kitchin and McArdle, 2016). Volume is a long-standing problem in spatial data analysis: an example is the Landsat programme in the 1970s which immediately generated more data than could be analysed. Although analysis and computing capabilities have grown enormously since the 1970s, so has the data volume because of geospatial technologies such as hyperspectral remote sensing, geosensor networks and location-aware technologies. Hidden in these vast volumes of data may be surprising and useful nuggets of information that generate new knowledge about the world (Miller and Han, 2009).

The variety of spatial data now available is also intriguing, but challenging: researchers are familiar with structured (quantitative data) in spatial data analytics but less familiar with unstructured (qualitative) data such as georeferenced text and imagery. They are also more comfortable with data generated from scientific instruments and professionals but less comfortable with data generated from consumer devices and citizens. The velocity (speed) of streaming data from the real word also creates challenges in acting on the data before the world changes and the data are stale, but also can allow new approaches to planning and management of complex human systems.

While progress with unprecedented spatial data volumes, variety and velocity are likely to lead to new insights, often overlooked is another V – *vanilla*. The ubiquity and ease of data collection change the nature of the analysis we can conduct in geography and other human and environmental sciences. The widespread deployment and adoption of georeferenced sensors facilitates naturalistic observation using the so-called digital exhaust generated as a byproduct of digitally enabled lifestyles, as well as volunteered and not-so-volunteered information. The collapsing cost and growing capabilities of data collection systems make it easier to conduct quasi-experiments such as instrumenting an environment and participants to observe reactions to an anticipated change in the real world (e.g. the impacts of a new public transit line on physical activity levels in a neighbourhood; see Brown et al., 2015). Also possible are retrospective experiments: since the data are collected on an ongoing basis, reactions to unanticipated events can be captured to reconstruct the dynamics that drove the event, as well as its cascading effects. Ubiquitous, ongoing data flows also allow us to capture spatio-temporal dynamics directly rather than inferring them from snapshots and at multiple temporal and spatial scales (Miller and Goodchild, 2014).

CHALLENGES

Spatial data analytics is becoming increasingly important: there is a spatial data revolution occurring across a wide range of scientific fields concerned with the Earth, its environments and people. It is also transforming business, cities and lifestyles, admittedly not always for the better. Some of these less-than-ideal outcomes should remind us that revolutions can fail. There are challenges across three domains that must be resolved for spatial data analytics to achieve their potential.

The first set of challenges concerns representation. Who are the people who use location-based services, social media and other location-aware technologies? Does the behaviour captured or shared by these services and devices describe their lives well, or is it a selective and biased insight? The second set of challenges concerns knowledge construction and delivery: how do we discover, confirm and express new knowledge from the torrents of data flowing from cities, societies and environments? Will this lead to better decisions in science, management and policy domains? And who has access to these data and tools and therefore gets to make (allegedly) better decisions? Finally, we must ask – will more information make us better off? Will the flood of new information and knowledge make the world a better place? This is not to suggest that spatial data analytics constitute a threat or are useless: clearly there are scientific advances to be had, and a world with substantial equity, sustainability and resilience challenges certainly needs help. Rather, what should give us pause is the indiscriminate flooding of information into all realms of human existence. There are domains and settings where allowing friction to slow things down may be beneficial. There are also times when spatial analysis should perhaps be avoided altogether.

WHO ARE THESE PEOPLE AND WHAT ARE THEY SHARING?

N ≠ ALL

In a sense, sampling is a reaction to the big data challenges of a previous era; when analysis was largely performed manually, dealing with large volumes of data was impractical and scientists developed methods for generalising small amounts of data to a larger unobserved population (Miller and Goodchild, 2014). Random sampling in all its forms (stratified, biased, spatial) maximises the likelihood that the limited sample we observe reflects the larger population we do not. This can work well, but it is fragile: it does not scale very well to subcategories, meaning that breaking the results down into finer-grained subcategories increases the possibility of erroneous predictions (Mayer-Schonberger and Cukier, 2013). It can compensate if the design of the sampling frame is known in advance to ensure randomness within these categories. But it does not work if the analysts want to re-use or repurpose the data for categories that they did not imagine when they conducted the sampling.

Data-driven science claims to sidestep sampling issues by providing the entire population under consideration, allowing data re-use and repurposing; this is often referred to as $N = all$: the sample *is* the population. However, in most cases this is false: it is rare when an analysis truly considers an entire population. Is what's missing different from what's not missing? Despite the ballyhoo about the closing of the digital divide, gaps still exist, particularly with respect to geography. For example, high-speed internet favours affluent neighbourhoods and cities, disfavouring poor neighbourhoods and rural areas, creating spatial and social basis in social media participation and civic hacking (Townsend, 2013). Haklay (2010) notes bias towards affluent and tourist areas in the OpenStreetMap project of volunteer-led mapping of the world.

Because of the $N = all$ myth, it is best to speak of sampled-versus-monitored populations rather than samples and populations. Using data from monitored populations creates a challenge that inverts the classic sampling problem. Rather than determining a question and collecting data to answer it, data are being generated and then it is being determined what questions can be answered (Miller and Goodchild, 2014).

TRUST BUT VERIFY

The variety of spatially and temporally referenced data available to analysts has increased dramatically. In addition to structured data such as quantitative measures, available too in digital form are unstructured data such as text, sound, imagery and video. There are challenges associated with processing unstructured data, such as extracting features from imagery; this is an active area of investigation that is making much progress. Also receiving much attention is integrated analysis of structured and unstructured data, with advances such as geovisual analytics, spatial decision support and mixed methods research (Andrienko et al., 2007; Johnson et al., 2007).

As Sui and Goodchild pointed out almost 15 years ago, GIS is not just a powerful geospatial calculator; GIS is also a powerful medium for communication (Sui and Goodchild, 2001).

More vexing are variations in data quality. We are witnessing a movement away from data generated from carefully calibrated scientific instruments and as part of a scientific model, and towards ad hoc collection using devices that are often designed for other purposes. A question we must ask with citizen science, social media and other volunteer information is: how do we know people are not mistaken – or maybe even lying (Goodchild, 2007; Flanagin and Metzger, 2008)? How do we verify asserted geographic facts? Syntactic geographic facts – the rules by which geographic reality or its measurements are constructed – are easier to check: examples include boundaries in cadastral databases or intersection geometry in transportation network databases. Less easily verified are semantic geographic facts, for example, the location of a geographic feature such as a lake or mountain peak (Goodchild and Li, 2012; Miller and Goodchild, 2014).

THE WORLD IS NOT ONE STAGE

The concept of a digital divide describes varying penetration rates of digital media into populations. Less recognised is a digital divide in people's lives between what they are sharing and what they are not sharing. Social media such as Facebook may have increasing penetration rates with respect to populations, but not necessarily similar penetration rates into people's lives (Miller and Goodchild, 2014). A half century ago, famed sociologist Irving Goffman, using a theatre as a metaphor for human behaviour, wrote about front-stage versus back-stage behaviours: behaviours meant for public consumption versus private consumption (Goffman, 1959). Social media may have blurred the line between these front-stage and back-stage behaviours for some people, but certainly not for everyone in all situations.

Is what people are sharing or trading a good surrogate for their lives, or is volunteered and traded information an incomplete portrayal? This of course depends on the questions being asked; it is difficult to answer without systematic inquiry. There is a need for careful comparisons between what can be understood from digital exhaust and other found data versus what requires careful (and expensive) sampling, attitudinal surveys and behavioural experiments. For example, mobility and social media data can tell us where and when people travel, but not why.

CAN SPATIAL DATA ANALYTICS TELL US ANYTHING NEW?

KNOWLEDGE: SOME ASSEMBLY REQUIRED

Knowledge construction requires the discovery, verification and expression of new understanding about some real-world phenomenon. Knowledge Discovery from Databases (KDD) (the traditional term that describes processes centred on 'data mining') could be viewed as an oxymoron: KDD and data mining do not discover knowledge – rather, they generate hypotheses that must be verified. The 19th-century scientist and philosopher C.S. Peirce termed this 'abductive reasoning': a form of inference that starts with data and ends with a hypothesis that describes the data. As Peirce noted, while deductive inference achieves what *must be* true, and inductive reasoning obtains what *is* true (for the data, at least), abductive reasoning only achieves what *could be* true: a highly tentative form of knowledge. Spatial knowledge discovered through data mining and exploratory techniques is the beginning of a scientific process, not the end. What is required are tools that support integrating casual reasoning using all three inferential modes (Gahegan, 2009).

NON-EXPRESSIBLE EXPLANATIONS

One strategy for leveraging big spatial data beyond the knowledge discovery phase is through data-driven modelling: allowing data rather than theory to generate models. Traditional model building is a top-down approach where a researcher articulates a theory or conceptual model in a mathematical or computational form, allowing it to be manipulated, estimated and tested against empirical data. Data-driven modelling is a bottom-up approach where the model form emerges from the data, using some automated or semi-automated search technique combined with some stated criterion for model success (such as goodness of fit).

Data-driven modelling presents several challenges. One challenge is that data-driven modelling takes away a powerful mechanism for constructing an explanatory model – theory. Theory tells us where and where not to look for explanations. A second challenge is that the data drive the form of the model, meaning there is no guarantee that the same model will result from a different dataset. With the same dataset, many different models could be generated that fit, meaning that slight alterations in the goodness-of-fit criterion or processes used to drive model selection can produce very different models (Fotheringham, 1998). A third challenge is the complexity of the resulting models. These often have non-intuitive functional forms and variable weightings. An early example is Openshaw's genetic algorithm-based automated modelling system for spatial interaction models; these generated models that defy explanation (Openshaw, 1988). But, if a modelling process generates something that cannot be expressed, has knowledge really been constructed (Miller and Goodchild, 2014)?

WHOSE ANALYTICS?

Much of the spatial and spatio-temporal data being generated through sensed environments and digital exhaust of spatial media are not free in either the 'free speech' or 'free beer' sense of that word. Rather, it is proprietary data generated through commercial applications and/or through participating in commercial relationships. This may entail a fundamental shift in science from the relatively open and transparent work of university researchers and government laboratories to less open and transparent realms.

Lazer et al. (2009) note that data-driven social science is emerging based on the capacity to collect and analyse massive amounts of data on individual and group behaviour. However, it is emerging in the private sector such as Google and Facebook, and in secretive government agencies, rather than in major journals in the social and economic sciences. The fear is that data-driven social science will become the exclusive domain of private companies, government agencies and a privileged set of academics working on non-transparent research that cannot be critiqued, published and replicated. This may not facilitate the advancement of science or serve the broader public interest in the accumulation and dissemination of knowledge (Miller, 2010); research gets black-boxed.

In his book on smart cities, Anthony Townsend (2013) notes that mobile technologies are transforming cities and regions into observatories where we can watch – in real time – how people move, how cities grow and shrink, economic dynamics and impacts on quality of life and health. But he also asks for whom we are constructing these smart infrastructures: citizens or corporations (Townsend, 2013)? To date, the development and deployment of smart infrastructures has largely been driven by corporations such as IBM, Siemens, Cisco and Intel. If smart cities are the observatories that facilitate new urban science and governance, we should be asking – who has the privilege of watching, for what purpose, and who is governing?

DOES THE WORLD NEED MORE SPATIAL INFORMATION?

FRICTION CAN BE A FRIEND

Spatio-temporal information, combined with mobility and communications technologies, are having a dramatic effect on the world. The development and deployment of space-adjusting technologies such as telephony, the internet and the automobile have generated a stunning degree of space–time convergence, or the apparent shrinking of space with respect to the time required to overcome space via mobility or communication, over the past two centuries. The world is shrinking, fragmenting and speeding up, with detrimental effects such as higher resource consumption, greater waste production, and increasing shearing forces between faster human systems and slower biological and physical systems (Janelle, 1969; Couclelis, 2000; Miller, 2013).

An often-stated goal of data analytics, and cyberinfrastructures such as smart cities, is efficiency: more efficient decision-making with more efficient outcomes. However, making systems more efficient will not necessarily make them more sustainable or resilient. Higher efficiency can increase rather than decrease resource consumption by lowering costs and market prices, generating higher demand. This is an effect known as *Jevons' paradox*, named after its 19th-century discoverer, the British economist William Stanley Jevons (Alcott, 2005). Jevons' paradox is evident in the explosion of mobility as a consequence of the collapse in transportation and communication costs over the past two centuries. While high mass mobility has individual benefits, it has collective harms such as congestion, non-renewable resource depletion, poor urban air quality, greenhouse gas generation and poor health outcomes such as obesity (Miller, 2013). These unintended consequences of the mobility explosion should give us pause as we flood cities and societies with spatial information via smart infrastructures.

Free-flowing data and instantaneous communication reduce friction by allowing people to sense quickly what is happening in a system and anticipate the future. However, the real problem facing humanity in the 21st century is not accelerating but rather slowing the flow of people and materials through cities and societies. A countervailing force to big data is what Townsend calls *slow data* (i.e. small data). Slow data are data collected sparingly and by design, not harvested opportunistically from data exhaust. Slow data can make the trade-offs between consumption and conservation explicit, dampening the positive feedback loop of efficiency and consumption (Townsend, 2013).

WHAT ARE YOU LOOKING AT?

It is difficult to write a chapter on big data and spatial data analytics without mentioning the potential violations of locational privacy (see Chapter 22). It is also difficult to discuss since the potential seems to be getting worse over time. Simply put, if you participate in the digital society, it is very easy to figure out who you are. For example, de Montjoye et al. (2013) found that four spatio-temporal data points from passive tracking of mobile phones are enough to identify 95% of users. Obfuscating these data through technical means such as locational masking is not a good solution since ensuring privacy requires dramatic degrading of the data (Kwan et al., 2004). Informed consent is the gold standard for university-based research, but the current opt-in model for most consumer applications services does not come close to meeting those standards. It is also difficult to imagine a contemporary lifestyle where a person opts out of digital tracking.

Locational privacy is a concern, not only as an intrinsic right but also as a source of backlash that will shut down spatial data analytics and data-driven science. The implications can range from embarrassment, to discrimination based on location, to what Jerome Dobson and Peter Fisher describe as geoslavery: a situation where one entity coercively or surreptitiously exerts control over the physical location of another entity, such as a person being intimidated to visit a mosque for daily prayers or community events since their locations may be monitored via their mobile phone (Dobson and Fisher, 2003). There is also the potential for monitoring pre-crimes, administering pre-punishments and predictive policing (Zedner, 2010): categorising and reacting to people based on likelihoods and correlations derived from spatial coincidence and social proximity rather than actual behaviour.

CONCLUSION

There is no question that big spatial data and spatial data analytics are a potential revolution in science, planning and management. These data sources and data-driven analytics are offering new insights into human behaviour, especially with respect to the small movements and mundane behaviours that sustain cities and societies but have been neglected by the past century of social and urban science (Batty, 2012). Although we may welcome the data-driven revolution, we certainly do not want to open the door to a data dictatorship (Mayer-Schonberger and Cukier, 2013). Data should not make our decisions for us: spatial data analytics and other data-driven science should support, not replace, decision-making by intelligent and sceptical scientists, planners and managers (Miller and Goodchild, 2014).

Avoiding the pitfalls of data hubris requires conscious and careful effort to ensure that our spatial analytical methods are transparent and replicable. We should also maintain respect for carefully sampled and measured authoritative data, using digital exhaust to seek unknowns rather than what can be known from authoritative data (Lazer et al., 2014). Spatial data analytics cannot be based solely on mining digital exhaust for correlation and coincidence. Data-driven science should seed (not grow) new theories and lead to actionable knowledge – these require thoughtful processes that are the hallmark of traditional science. As Michael Batty says, '[t]here's all this new stuff, but the old questions are still here and they've not been answered' (quoted in Townsend, 2013: 315).

REFERENCES

Alcott, B. (2005) 'Jevons' paradox', *Ecological Economics*, 54: 9–21.

Andrienko, G., Andrienko, N., Jankowski, P., Keim, D., Kraak, M.-J., MacEachren, A. and Wrobela, S. (2007) 'Geovisual analytics for spatial decision support: setting the research agenda', *International Journal of Geographical Information Science*, 21: 839–57.

Batty, M. (2012) 'Smart Cities, Big Data', *Environment and Planning B*, 39: 191–3.

Brown, B.B., Werner, C.M., Tribby, C.P., Miller, H.J. and Smith, K.R. (2015) 'Transit use, physical activity, and body mass index changes: objective measures associated with complete street light rail construction', *American Journal of Public Health*, 105 (7): 1468–74.

Couclelis, H. (2000) 'From sustainable transportation to sustainable accessibility: can we avoid a new tragedy of the commons?', in D.G. Janelle and D. Hodge (eds), *Information, Place and Cyberspace: Issues in Accessibility*. Berlin: Springer. pp. 341–56.

de Montjoye, Y.-A., Hidalgo, C.A., Verleysen, M. and Blondel, V.D. (2013) 'Unique in the crowd: the privacy bounds of human mobility', *Nature Scientific Reports*, 3: 1376.

Dobson, J.E. and Fisher, P.F. (2003) 'Geoslavery', *IEEE Technology and Society Magazine*, Spring: 47–52.

Dumbill, E. (2012) 'What is big data? An introduction to the big data landscape', O'Reilly. Available at: http://strata.oreilly.com/2012/01/what-is-big-data.html (accessed 26 July 2016).

Flanagin, A.J. and Metzger, M.J. (2008) 'The credibility of volunteered geographic information', *GeoJournal*, 72: 137–48.

Fotheringham, A.S. (1998) 'Trends in quantitative methods II: stressing the computational', *Progress in Human Geography*, 22: 283–92.

Gahegan, M. (2009) 'Visual exploration and explanation in geography: analysis with light', in H.J. Miller and J. Han (eds), *Geographic Data Mining and Knowledge Discovery*, 2nd edition. London: CRC Press. pp. 291–324.

Goffman, E. (1959) *The Presentation of Self in Everyday Life*. New York: Anchor Books.

Goodchild, M.F. (2007) 'Citizens as sensors: the world of volunteered geography', *GeoJournal*, 69: 211–21.

Goodchild, M.F. and Li, L. (2012) 'Assuring the quality of volunteered geographic information', *Spatial Statistics*, 1: 110–20.

Haklay, M. (2010) 'How good is volunteered geographical information? A comparative study of OpenStreetMap and Ordnance Survey datasets', *Environment and Planning B: Planning and Design*, 37: 682–703.

Janelle, D.G. (1969) 'Spatial organization: a model and concept', *Annals of the Association of American Geographers*, 59: 348–64.

Johnson, R.B., Onwuegbuzie, A.J. and Turner, L.A. (2007) 'Toward a definition of mixed methods research', *Journal of Mixed Methods Research*, 1: 112–33.

Kitchin, R. and McArdle, G. (2016) 'What makes big data, big data? Exploring the ontological characteristics of 26 datasets', *Big Data and Society*, 3: 1–10.

Kwan, M.-P., Casas, I. and Schmitz, B. (2004) 'Protection of geoprivacy and accuracy of spatial information: how effective are geographical masks?', *Cartographica*, 39 (2): 15–28.

Lazer, D., Pentland, A., Adamic, L., Aral, S., Barabási, A.-L., Brewer, D., Christakis, N., Contractor, N., Fowler, J., Gutmann, M., Jebara, T., King, G., Macy, M., Roy, D. and Van Alstyne, M. (2009) 'Computational social science', *Science*, 323 (5915): 721–3.

Lazer, D., Kennedy, R., King, G. and Vespignani, A. (2014) 'The parable of Google Flu: traps in Big Data analysis', *Science*, 343 (6176): 1203–5.

Mayer-Schonberger, V. and Cukier, K. (2013) *Big Data: A Revolution that Will Transform How We Live, Work, and Think*. London: John Murray.

Miller, H.J. (2010) 'The data avalanche is here. Shouldn't we be digging?', *Journal of Regional Science*, 50: 181–201.

Miller, H.J. (2013) 'Beyond sharing: cultivating cooperative transportation systems through geographic information science', *Journal of Transport Geography*, 31: 296–308.

Miller, H.J. and Goodchild, M.F. (2014) 'Data-driven geography', *GeoJournal* 80 (4): 449–61.

Miller, H.J. and Han, J. (2009) *Geographic Data Mining and Knowledge Discovery*, 2nd edition. London: CRC Press.

Openshaw, S. (1988) 'Building an automated modeling system to explore a universe of spatial interaction models', *Geographical Analysis*, 20: 31–46.

Sui, D.Z. and Goodchild, M.F. (2001) 'GIS as media?', *International Journal of Geographical Information Science*, 15: 387–90.

Townsend, A. (2013) *Smart Cities: Big Data, Civic Hackers, and the Quest for a New Utopia*. New York: Norton.

Zedner, L. (2010) 'Pre-crime and pre-punishment: a health warning', *Criminal Justice Matters*, 81: 24–5.

15

LEGAL RIGHTS AND SPATIAL MEDIA

TERESA SCASSA

INTRODUCTION

Despite dramatic technological changes that have revolutionised how data are collected, compiled and used — and by whom — the law regarding the ownership of compilations of data has remained fairly static since the 1990s. What has changed, however, are the number and kind of users of spatial data, the new technologies that expand categories of spatial data, and the way many governments now approach the licensing of their stores of spatial data. This chapter considers the interrelationship between claims to property rights in data, and rights to access and use those data in a rapidly changing technological environment. In doing so, it considers law not simply in terms of legislation and case law, but also in terms of the norms that evolve within communities of practice. The chapter begins with a discussion of the legal status of claims to property rights in data and their potential weaknesses and contingency. It then considers the role of norms within communities of practice around the re-use of data and the licensing of government data. Practices around the licensing of data reflect a sometimes perilous balance between rights, expectations and the relative power of those involved in the transactions.

LEGAL RIGHTS AND SPATIAL DATA

The collection or generation of spatial data is often the result of a significant investment of time, money and labour. As a result, compilations of spatial data are routinely treated by their compilers as a form of intellectual property (IP). The propertisation of data allows an owner to construct fences around the data in order to exclude unauthorised users and to control authenticity and authority. Where an owner chooses to permit access or use, they can do so through a series of contractual terms that may impose a broad range of conditions and limitations. Although a combination of technological barriers and contractual terms can be used to restrict access and use, even without an underlying property right, IP rights provide a level of control that extends beyond any technological or contractual restrictions. This is because IP rights encompass a right to exclude that is not dependent upon any pre-existing relationships. Contracts, by contrast, require two or more parties to agree to certain terms of access – only parties privy to a contract are bound by it. In such a context, it is not surprising that property rights are asserted over data, even where their existence may be questionable or limited in nature.

Copyright principles include a fact–expression dichotomy that draws a line between what may be protected (original expression) and what may not (facts). In old-tech terms, it was relatively easy to separate fact from expression. For example, spatial media have always involved both geospatial data and representations or expressions of those data. In the case of old-fashioned, paper-based maps, the cartographer's graphic expression of geographical facts was clearly distinguishable from the underlying facts. Maps can be protected by copyright law as artistic works in jurisdictions such as the UK and Canada (Judge and Scassa, 2010) or as compilations of facts in the US (Martino, 2006). In either case, however, the legal protection available does not extend to the underlying facts. While each thematic mapmaker is entitled to protection for his or her original expression of geographical facts, the public policy underlying copyright law prevents anyone from obtaining a monopoly over the ability to represent those facts in a map.

The simple dichotomy in map cases is between fact and its expression, reflecting the distinction between what can and cannot be 'owned'. Distinguishing between fact and expression is rendered more complex by advances in technology. In the realm of spatial media, therefore, the dichotomy between data and their expression is complicated by layered expressions (databases and software underlying visual digital representations, for example); by multimedia expressions capable of different IP owners (images, text, video); and by the capacity to mash up data from multiple proprietary sources. The easier it is to share, manipulate, combine and represent data, the more problematic it becomes not only to separate fact from expression, but also to navigate complex webs of ownership.

As the distinction between data and their expression became more complex, the law around the protection of compilations of data grew in importance. Historically, compilations of facts (or data) were protected under copyright law. However, copyright law protects only original expression, and such expression is considered to lie in the manner in which the data are selected or arranged. *Feist v. Rural Telephone Co.* (1991) sent shockwaves through the database industry when the US Supreme Court firmly ruled out 'sweat of the brow' (effort or investment) as providing a basis for a finding of originality in the selection or arrangement of data. It also found that the compilation of data at issue in that case – a telephone directory – lacked originality in its selection (all subscribers who had not opted to have unlisted numbers) and in its arrangement (in alphabetical order) of the data. Further, the court reminded the legal world that facts were in the public domain. Copyright protection could extend only to the original selection or arrangement of these facts; another user who made their own selection and arrangement of the same facts would not be taking any of the original creator's expression. The decision created two dilemmas for producers of compilations of data. The first was the legal proposition that not all compilations of data were capable of protection under copyright law; the second was that the entitlement to such protection – and its scope – would, in many cases, remain a matter of speculation absent a binding court decision (Green, 2009).

The European response to post-*Feist* calls from the database industry for greater legal certainty was the EU *Database Directive* (EC, 1996). According to the *Directive*, a database is defined as 'a collection of independent works, data or other materials arranged in a systematic or methodical way and individually accessible by electronic or other means'. Article 7 provides for the protection of databases in which 'there has been qualitatively and/or quantitatively a substantial investment in either the obtaining, verification or presentation of the contents'. The database right entitles the owner 'to prevent extraction and/or re-utilization of the whole or of a substantial part, evaluated qualitatively and/or quantitatively, of the contents of that database' (art. 7). Protection under this directive is for a 15-year term (much lower than the copyright term of life of the author plus 70 years). However, a new 15-year term of protection is available each time the database is substantially revised. Database rights are considerably more certain than copyright in terms of both their scope and subsistence.

The split between 'old' and 'new' world approaches to rights in databases poses challenges for re-use since the legal rights underlying any data licence may vary significantly in scope and certainty from one jurisdiction to another. This is not the only important difference. Much spatial data is collected by governments, a significant majority of which exercise IP rights over their compilations of data. The United States is an important exception to this general rule. The US *Copyright Act* provides that copyright protection is not available for 'any work of the United States Government' (§105). Such works – which include compilations of data – are treated as falling within the public domain. However, the exception to copyright protection applies only to the Federal government; state and municipal governments in the US can and do assert copyright over their works, including data. In countries such as the UK, Canada, Australia and New Zealand, 'Crown copyright' forms the basis for government claims (at all levels of government) to copyright in government works. These differences in protection available for government compilations of data can lead to different attitudes towards the use of such data within the user community. Among legally unsophisticated users, the understanding that US government data is in the public domain can spill over into assumptions that all government data – both domestically and internationally – are, or should be – in the public domain.

Spatial media include more than just databases. Copyright law also protects software, and it will protect text, photographs, videos and other multimedia content that is present in some online spatial media such as cybercartography (Scassa and Taylor, 2014). This can lead to a complex layering of rights: a database of satellite images may (or may not) have copyright or database protection for the database as a whole as well as copyright protection for each individual image. In some cases, the rights may be contingent and contestable. For example, satellite images or images captured by ocean floor cameras or traffic cameras (to give a few examples) may be treated differently from one jurisdiction to the next (e.g. as compilations of data in the US, and as photographs in the UK) (Burk, 2007). Where such images are considered to be photographs, they may lack the authorship or originality necessary to meet the threshold for originality under copyright law (Hughes, 2012). As compilations of data, they may lack originality, since the selection or arrangement of the data they embody does not derive from a human author (Perry and Margoni, 2010). It is also possible that courts in different jurisdictions will answer questions of this kind in different ways.

Issues as to rights in data may be further complicated by disputes over ownership of those rights. For example, patent law can protect functional aspects of software, as well as technologies for gathering and communicating spatial data. This added proprietary layer may raise interesting issues around the ownership of data that are gathered using the patented technologies. For example, a municipal transit authority that contracts for automated vehicle location services from a private sector company, to collect global positioning system (GPS) data regarding its transit vehicles, may find itself in a dispute over ownership of the resultant data (Scassa, 2014). While the patent rights would not extend to the data, the argument might be that data generated using the patented technology belong to the owner of that technology. Issues like this can be particularly important where the transit authority seeks to make the data open to developers as part of an open data programme (see Chapter 9), and where the developers might create apps that compete with services offered by the private sector corporation (Scassa, 2014). This is an important issue, since as so-called 'smart' technologies evolve, the collection by governments of live-streaming georeferenced information will become commonplace (see Chapter 11). The ability of governments to make such data available as open data may depend upon whether IP rights are located with the government or with the private sector partner.

New types of data may also raise interesting issues regarding the subsistence of copyright. For example, there is some controversy as to whether predictive data are excluded from copyright protection under the fact/expression dichotomy – since predictions are arguably not factual (Green, 2009; Thomas, 2011). The answer may depend not only on how the prediction is derived (i.e. the algorithm, software code, etc.) but on how it is expressed – and it may vary from one jurisdiction to another.

The key point is that there is considerable uncertainty and contingency when it comes to rights in spatial media and their underlying components. These uncertainties chiefly arise:

- as between different countries;

- as to the subsistence of rights in some subject matter;

- as to the subsistence of rights in particular works or compilations of data;

- as to the scope of any rights that might exist in a particular work or compilation.

The next section explores how these legal uncertainties may be addressed within communities of practice.

SPATIAL DATA: COMMUNITIES OF PRACTICE

In the previous section, we considered the extent to which differences between jurisdictions over the source and scope of IP rights – combined with the need in some circumstances for a case-by-case assessment of the subsistence and scope of rights – can create uncertainty. The differences in type and level of protection, and from one jurisdiction to the next, become more acute where developers seek to use data from multiple sources and of multiple types in complex multimedia projects. Because legal uncertainty can be a barrier to innovation, communities of practice often develop within particular sectors to reduce these uncertainties. These communities develop customary practices that may be reinforced by contracts or licences that reflect settled expectations around IP rights.

As Murray et al. (2014: 1) observe, 'effects attributed to IP statute and case law are often, in fact, results of cultural, professional, economic, and ideological circumstances in which IP law is invoked or imagined occasionally, opportunistically, or instrumentally'. One of the characteristics of a data 'ecosystem' (a set of creators and users of a particular type of data) is the emergence of a working consensus around the boundaries of property rights in data. This may occur even absent of any case law that establishes or delimits rights. Murray et al. (2014: 2) define IP law as 'the complex interactions between statute and case law (what lawyers would call "IP law") and the ways it is understood or mobilized as a symbol or discourse'. It is this version of IP law that is significant within communities of practice.

The practices among users of spatial data have evolved along with the technologies used in the gathering, processing and representation of those data. For example, commercial users might seek to license the data they need and to negotiate for an acceptable price and conditions. All actors within a given industry adapt to and accept these for-fee licences as a cost of doing business. Yet the practice can be vulnerable to anyone who chooses not to play by the rules since litigation may reduce or eliminate the licensor's rights. This vulnerability is well-illustrated by *Feist*, discussed earlier. At the time of this litigation, it was a well-established practice in the telephone directory industry in the US for compilers of regional telephone directories to license data from the relevant telephone companies in order to create their directories. Rural Telephone's refusal to license its data to Feist led to litigation and ultimately to a decision that entirely eliminated the need for directory data to be licensed (Landes and Posner, 2003).

One response to the legal uncertainty over copyright protection for compilations of data has been simply for database owners to assert rights regardless of their real scope or extent. The assertion of rights backed by threats to enforce these rights if necessary is often sufficient to achieve an adequate level of protection (McBrayer, 2005). Users of the data must balance the cost of licensing against the costs and uncertainty of litigation. It is often more efficient to pay for use rather than to risk litigation. Within an industry, this can result in settled expectations and practices that provide normative structure and reduce concerns over uncertainty. However, it is a normative structure that is insensitive to disparities in economic power. The fewer resources a party has, the lower its ability to resist the assertion of weak or even non-existent intellectual property rights. In the US, frustrations over the impact of the assertion of weak or unfounded rights claims have given rise to the evolution of the doctrine of copyright misuse. Such practices can pose barriers to market entry, may reflect anticompetitive behaviour and can create risks that some will be deliberately excluded from participation (Judge, 2004).

Battles over open government data mark an important shift within communities of practice around data. At a time when only major corporations had the resources to create downstream products using government data, it made more sense to commercially license these data. As even individuals and small start-ups became able to derive value from data, for-fee licensing posed a significant barrier both to entry into markets and to innovation. The pressure for more equitable access to government data – for both commercial and non-commercial purposes – led to the open government data movement, which has significant overlaps with the drive to 'open' government spatial data (Lauriault and Kitchin, 2014; see Chapters 9 and 16). Where a government implements 'open data', it makes government datasets available to the public in re-usable formats under an open licence (Kitchin, 2014). Respect for and compliance with the open licence amounts to acceptance of the underlying claims to intellectual property rights in the data – notwithstanding the fact that these claims may well be weak or contingent. Nevertheless, it is more practical and efficient to accept these claims in exchange for licence terms that permit re-use of the data under relatively few conditions, none of which, in theory at least, should be onerous. It should be noted that although it may reset the norms of communities of practice with respect to government data, open data licensing (see Chapter 9) also contributes to the reinforcement of established views about the subsistence and scope of IP rights in compilations of data within these communities of practice.

LICENSING OF GOVERNMENT SPATIAL DATA

There have been significant debates over access to and use of spatial data resources, particularly those that are in the hands of governments. Many (though not all) governments assert IP rights over their data, and do so as a means of control. Masser (1998: 16–17), for example, refers to governments as custodians of public information. The open data movement has pushed governments to relinquish this control, resulting in the release of government datasets under licences that contain few if any restrictions. In the transition from closed to open data, governments have had to abandon or alter many of the public policy justifications that underpinned their exercise of control over data. These justifications have included: ensuring accuracy and authoritativeness (Judge, 2005), limiting risk and liability, and allowing for cost recovery or even profit generation (Masser, 1998). Open government data is thus more than simply a decision to stimulate innovation and re-use of government data; it reflects a major shift in policy direction and a re-negotiation of the relationship between data users and data providers.

The growing embrace of open government data (see Chapter 18) by different national, regional or local governments, combined with the vast diversity of types of data and their growing complexity (both in terms of the data themselves and the means by which they can be analysed, combined and manipulated), have all increased the legal complexities around ownership and re-use. One issue is so-called legal inter-operability – the idea that if the open licences attached to different datasets from different sources are not compatible this will create legal uncertainty about the ability to combine data from these different sources in new and innovative ways (Mewhort, 2012). At the same time, some governments – perhaps increasingly concerned about their legal obligations to maintain the privacy of the personal information they collect in the light of big data analytics that facilitate re-identification – have chosen to include terms in open licences that limit the right to re-use only to data that do not implicate privacy rights (National Archives, 2015; Government of Canada, 2015). Some licences may also exclude from their scope any data that may have been provided by third parties, including data which governments are legally obliged to keep confidential (National Archives, 2015; Government of Canada, 2015). Some open government licences additionally contain attribution requirements and limitations on any claims of endorsement by government of downstream data products or services (Government of France, 2011). The insertion of such clauses contributes to issues around the compatibility of open licences and may create uncertainty as to the scope of the licence granted. The situation is exacerbated by the growing number of potential 'works' that may be created using data from different sources, the proliferation of potential rights-holders, the diversity of potentially applicable legal regimes, and the number of different licences that may be at issue in any given project or application.

Technological change may pose other licensing challenges. Providing access to live-streamed data such as real-time GPS data involves more than making datasets available under an open licence – the nature of the data means that developers will need frequent, real-time access to the data via an application programming interface (API). Research in the transit data context shows that even where the data themselves are available under an open licence, the contractual terms of use for the API place additional restrictions on both access to and use of the data (Scassa and Diebel, 2016). In some cases, API terms of use reintroduce many of the restrictive clauses that have disappeared from open licences. Yet in addition real-time data introduce new resource and liability issues for governments that may also pose challenges to the ability to open the data under the same terms as static data. The shifting nature of data as technology advances thus adds another layer of legal complexity to data re-use in this context.

CONCLUSIONS

The rapidly evolving data landscape, and the ways in which the data revolution (Kitchin, 2014) is changing both the delivery of government services and the kinds and quantity of data generated by these services are once again poised to transform how ownership of and access to data is negotiated between data owners and data users. In the context of 'public' services, ownership and control issues will be complicated by the presence of private-sector companies who partner in the collection and generation of data (i.e. smart cities, geodemographics). The growing importance of live-streamed data and predictive data, to give two examples, disrupts conventional approaches both to property rights and to contractual terms of access and use.

Already existing communities of practice may evolve and adapt to these changing circumstances. These communities now include open data activists, developers and small entrepreneurs, and the changing demographics of the community of practice may influence their evolution. It is likely that development of legal norms in this area may be more impacted by the changing attitudes of the community members than by any process of law reform or adjudication.

REFERENCES

Burk, D.L. (2007) 'Intellectual property in the context of e-Science', *Journal of Computer-Mediated Communication*, 12: 600–17.

EC, European Database Directive (1996) '96/9/EC of the European Parliament and of the Council of the European Union of 11 March 1996 on the legal protection of databases', O.J. L 77/20.

Feist Publications, Inc. v. Rural Telephone Service Co., 499 U.S. 340, 111 S.Ct. 1282 (1991).

Government of Canada (2015) 'Open Government Licence – Canada, Version 2.0'. Available at: http://open.canada.ca/en/open-government-licence-canada (accessed 26 July 2016).

Government of France (2011) 'Licence ouverte'. Available at: https://www.etalab.gouv.fr/en/licence-ouverte-open-licence (accessed 26 July 2016).

Green, M.S. (2009) 'Two fallacies about copyrighting factual compilations', in R.F. Brauneis (ed.), *Intellectual Property Protection of Fact-based Works: Copyright and Its Alternatives*. Cheltenham: Edward Elgar. pp. 109–32.

Hughes, J. (2012) 'The photographer's copyright – photograph as art, photograph as database', *Harvard Journal of Law and Technology*, 25: 339–428.

Judge, E.F. (2005) 'Crown copyright and copyright reform in Canada', in M. Geist (ed.), *In the Public Interest: The Future of Canadian Copyright Law*. Toronto: Irwin Law. pp. 551–94.

Judge, E.F. and Scassa, T. (2010) 'Intellectual property and the licensing of Canadian government geospatial data: an examination of geoconnections' recommendations for best practices and template licences', *Canadian Geographer*, 54 (3): 366–74.

Judge, K. (2004) 'Rethinking copyright misuse', *Stanford Law Review*, 57: 901–52.

Kitchin, R. (2014) *The Data Revolution: Big Data, Open Data, Data Infrastructures and their Consequences*. London: Sage.

Landes, W.M. and Posner, R.A. (2003) *The Economic Structure of Intellectual Property Law*. Cambridge, MA: Harvard University Press.

Lauriault, T.L. and Kitchin, R. (2014) 'A genealogy of data assemblages: tracing the geospatial open access and open data movements in Canada', presentation at AAG Annual Meeting, Tampa Bay, FL.

McBrayer, L. (2005) 'The DirectTV Cases: applying anti-SLAPP laws to copyright protection cease-and-desist letters', *Berkeley Technology Law Journal*, 20: 602–24.

Martino, P. (2006) 'Clarifying copyrightability in databases', *The Georgetown Journal of Law and Public Policy*, 4: 557–94.

Masser, I. (1998) *Governments and Geographic Information*. London: Taylor & Francis.

Mewhort, K. (2012) 'Creative commons licences: options for Canadian open data providers', Sauelson-Glushko Canadian Internet Policy and Public Interest Clinic, 1 June.

Available at: https://www.cippic.ca/sites/default/files/Creative20%Commons20% Licenses20%–20%Options20%for20%Canadian20%Open20%Data20%Providers. pdf (accessed 26 July 2016).

Murray, L., Piper, S.T. and Robertson, K. (2014) *Putting Intellectual Property in Its Place: Rights Discourses, Creative Labour, and the Everyday*. Oxford: Oxford University Press.

National Archives (UK) (2015) 'Open government licence for public sector information, version 3'. Available at: http://www.nationalarchives.gov.uk/doc/open-government-licence/version/3/ (accessed 26 July 2016).

Perry, M. and Margoni, T. (2010) 'From music tracks to Google maps: who owns computer-generated works?', *Computer Law and Security Review*, 26 (6): 621–9.

Scassa, T. (2014) 'Public transit data through an intellectual property lens: lessons about open data', *Fordham Urban Law Journal*, 41: 1759–810.

Scassa, T. and Diebel, A. (2015) 'Open or closed? Licensing real-time GPS data', *Journal of e-Democracy*, 8 (3): forth coming.

Scassa, T. and Taylor, D.R.F. (2014) 'Intellectual property and geospatial information: some challenges', *WIPO Journal*, 6 (1): 79–88.

Thomas, L.W. (2011) 'Legal Research Digest 37: legal arrangements for use and control of real-time data', Washington, DC: Transportation Research Board of the National Academies. Available at: http://www.trb.org/Publications/Blurbs/165626. aspx (accessed 26 July 2016).

US *Copyright Act*, 17 U.S.C. §§ 101–80.

PART 3:
THE CONSEQUENCES
OF SPATIAL MEDIA

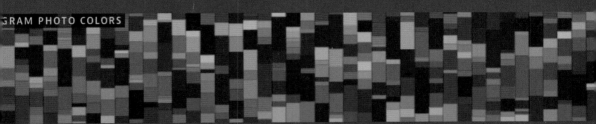

16

SPATIAL KNOWLEDGE AND BEHAVIOUR

LEIGHTON EVANS AND *SUNG-YUEH PERNG*

INTRODUCTION

Using spatial media to understand, navigate and act on the world is an integral element in being a connected, technology-using individual. Apple users discovered just how critical spatial media are to their everyday life when updating their mobile devices to the iOS6 operating system on 19 September 2012. Following the long process, users discovered that the Google maps application had been replaced with Apple's own maps application. The new application, built on OpenStreetMap, brought much derision due to its curious inaccuracies, such as labelling Berlin as 'Schoeneiche' (Butcher, 2012). The changes brought consternation and discontent; Apple swiftly allowed Google Maps to be reinstalled as an app on iOS devices in December 2012. Despite these issues, evidence suggests that many users continued to use Apple's app following improvements (Arthur, 2013), underlining the importance of mapping apps for smart phone users. The disruption of spatial media services caused a major public relations incident for Apple, but it also created angst for media users as these applications have become critical for the continually connected mobile media user and have disrupted their perceived ability to cope in the everyday world. The behaviour of using spatial media and the epistemologies that this engenders are therefore critical features of our data- and information-infused world. This chapter investigates some of the emerging issues and theoretical approaches related to this behavioural and epistemological shift.

EMERGING SOCIAL AND SPATIAL ISSUES

Spatial media have made their way into our everyday life at an expedited speed, quickly supplementing our spatial knowledge, and reshaping behaviour. The changing relationship between individuals and the spatial media they use is a starting point to examine the effect of spatial media. Rather than providing decontextualised information that has little or no relevance to location, current everyday spatial media (e.g. location-based social networks or augmented-reality browsers) allow users to focus on knowing about where they are and provide location-sensitive feeds of information (McCulloch, 2006). Individual users are not atomised actors; spatial media acts as a kind of 'me-dia' (Merrin, 2014: 1) where the device focuses upon the individual in a hyper-localised manner, providing and producing specific and hyper-relevant information as a form of horizontal, peer-to-peer, mediated interpersonal communication that affects epistemologies of space and place.

This, however, is not to say location-sensitive technology is the prerequisite for such peer-to-peer communication. Devices can become 'spatial' through social practices even when they are not explicitly designed for that purpose. Mobile phone conversations, as often overheard in public places, become spatial in that 'the giving of a geographical formulation' is part and parcel of 'the opening sequence of a phone call' (Laurier, 2001: 485). Likewise they act as a spatial tool for the coordination of where to meet, wherein meetings can be arranged on the fly with last-minute changes or just-in-time updates (Ling and Haddon, 2003; see Chapter 1).

The possibility and affordance of this instant interactivity with others who surround them can transform what it means to participate and become 'the public', and indeed the meaning of 'publics'. Spatial media services mediate conceptions of space and geography while contributing to changes in understandings of participation in public life for users (Campbell and Kwak, 2011; Gordon et al., 2013). Humphreys' research (2005, 2008, 2010) suggests people map their understanding of common social rules and dilemmas onto new technologies, and over time this creates a new social landscape. Spatial media use, therefore, becomes assimilated into common social practices. Humphreys (2008) researched the LBSN 'Dodgeball' (now defunct) and found that using the application influenced the way participants in the study experienced public space and the social relations therein. 'Dodgeball' use facilitated the creation of 'third spaces', which are dynamic and iterant forms of social place, distinct from public or private space through their mediation by mobile media. More generally, Sheller (2004: 39) argues that spatial media use necessitates a rethinking of the term 'publics'. Mechanisms that enable and constrain different ways private citizens engage in public participation are emerging from intersecting mobilities of information, technology and people that reshape the spatial and temporal patterning of publicity and privacy. Such reshaping of spatiality has encouraged the forming of 'mobile publics', a new way of momentary 'gelling' of identities and actions by enabling certain contexts, motivations and relationships to converge across spaces and scales. Similarly, Humphreys (2010) returned to these themes by arguing that mobile social networking has the potential to transform the ways that people come together and interact in public space, and allows new kinds of information to flow into public spaces, which in turn changes our social behaviours and understanding of those spaces. Flash mobs facilitated by mobile media, where disparate groups and individuals converge on a space at a set time for a shared social experience, would be an example of this. What this body of research emphasises is that using spatial media changes the way users actively participate in everyday life. The co-presence of others through spatial media (particularly locative social media such as Foursquare) and the ability to draw on this co-presence change the way users inhabit space and behave while using the technology in a place.

Another critical issue concerns embodied experiences in perpetually mediated worlds. Farman (2012) explains the relationship between user and spatial media through the prism of embedded cognition, where embodiment and space are co-constitutive for understanding of places. Spatial media here actively reconfigure the way that users are embodied in space (i.e. are physically in that space) through changes in their behaviour and changes in how they seek out information about place. This approach positions the user as an active part of the mediation of the world. Farman argues that mobile technologies are reconfiguring the ways in which users can embody space and locate themselves in digital and virtual spaces simulta-neously. Consider the flash mob: in that activity, users inhabit a physical space while performing an activity that has been coordinated (and often is being streamed or tweeted directly to) in a virtual space. Knowledge of spaces therefore becomes a complex interdependent process where presence, practices of use, software and data are all intricately linked in producing knowledge and understanding.

Evans (2015) develops a similar approach, arguing that the understanding and knowledge of space and place when using spatial media is dependent upon the mood or orientation of the user. If the user is orientated towards the world in a manner that seeks to understand the place as meaningful, then the practices of using spatial media can facilitate this and aid familiarity in unfamiliar spaces. As Frith (2012) puts it, spatial media afford the possibility of a personal database city where the subjective experience of places is both coded into databases and fed back to users, making the device- and location-based service a critical aspect of the subjective experience of place. Following Evans (2015), such knowledge is contingent on the intentionality or orientation of the user towards place and the spatial media being used.

These various engagements and forms of knowledge emerging from the use of spatial media also become sources of formulating privacy issues in our hyper-mediated worlds. Location-sensitive media post unique challenges to privacy because users are simultaneously in social, embodied, virtual and physical 'places' (see Chapter 22). The social norms and expectations associated with physical proximity suddenly lose their clarity: the co-presence in a place can happen with little previous personal relationships; and knowledge about private places no longer presupposes or guarantees exclusion of the public (Licoppe and Inada, 2008).

Hjorth's (2013) two-year ethnography of mobile media users in Seoul concen-trated on the renegotiation of privacy and place in light of mobile media use. Hjorth focused on the ways that users grappled with spatial media and how using appli-cations to move through and navigate urban space includes a continual trade-off between the advantages of use and privacy concerns created by surrendering data to companies. Spatial media use therefore does not exist in a vacuum with regards to privacy. Users are aware of the production and sharing of location-based data by their activity, and this is a factor in use. These entangled relationships between places, spatial media and users thus call for careful consideration of the often fluid and dynamically changing contextualisation of privacy and its consequences and dangers to civil liberty (Palen and Dourish, 2003; Büscher et al., 2013).

Another stream of practices seeks to explore opportunities for civic collective agency through incorporating locative media technologies into interdisciplinary collaboration between natural and social scientists, artist-designers and publics. Such collaboration seeks to provoke rethinking and reshaping normalised knowledge and experiences of cities, and their dramatically changing physical, atmospheric or climatic environments (Hemment et al., 2011). Similarly, the totalisation of socio-technological experiences in urban spaces can be challenged by understanding the decentralised, distributed and relational perfomativities that are situated in those spaces. Through engaging with mobile technologies, including phones and global positioning systems (GPS), locative game players produce their own memories and experiences of the space that are indeterminate and, ongoing, and resist easy categorisation (Galloway and Ward, 2006). GPS technologies can be further reappropriated in artistic and everyday performances in ways that record the movements of individuals and simultaneously connect them with those of their friends and loved ones (Southern, 2012). These data together form a web of each other's locations and the paths they have undertaken, and display and articulate 'side-by-side' views of the life and travel among themselves as their life events unfold. These performances demonstrate how spatial relations are configured in the ways people, locations and GPS technologies relate to one another, and thus urge us to rethink the assumption of an all-knowing position for knowledge production that is prevalent in modern (spatial) sciences and technologies.

The most salient feature of all these practices is the seemingly asynchronous relationship between hyper-local activities that draw on and link back to distant databanks. Gordon and de Souza e Silva (2011) use the concept of networked locality to emphasise how particular usage of networked devices, and the information they can provide from de-localised storage, can increase nearness to places rather than increase distance from spaces and people. The provision of locally and personally specific information on place in fact can make us feel more connected to the place we are in – even if those data are stored in a databank thousands of miles away. It is the context of the information and how this contextual information is received and interpreted by the user that is important. Shepard (2011) thus argues that the 'data-clouds of the 21st century' (accessed through spatial media services) increasingly shape experience of physical space; the data collected in distant databanks play a key role in how we understand the space we physically are in when we use spatial media.

THEORETICAL APPROACHES TO THESE ISSUES

Theory and research on spatial media have looked for more refined views on the relationships between technologies, places and structures that go beyond any form of determinism. Silverstone (1994) argues that the places where technology's effects are conditioned by social institutions had not been sufficiently conceptualised in previous theorisations (most notably McLuhan and Meyrowitz) and this remains the case. The extent to which our behaviour and understanding of place can be attributed to structural and technological framing is a question requiring careful consideration. Williams (2003) contends that we have to think of determination not as a single force, or a single abstraction of either social or technological forces, but as a process in which real determining factors set limits and exert pressures. In other words, the technologies contribute as a factor in how the world is viewed, rather than dominating it. As such, spatial media, as a factor in understanding the world through its usage, should be considered as a factor among others in the shaping of the world. Software, as a contemporary example, has become a mature part of societal formations, being a constituent element of daily life (or at least is in a phase where generations are now born into using it), and social science needs to make understood a range of associations and interpretations of software (Fuller, 2008: 18). When one views the world with the assistance of computational devices, software or code, one is orientated towards the world in a way that assumes it has already been mapped, classified and digitised. Space and place are co-constructed through computational devices that offer this worldview back through a plethora of computational mediators, such as mobile phones, car navigation systems or handheld computers that are our 'familiar' (Berry, 2011: 137).

Various approaches have subsequently sought to offer nuanced ways of theorising the implications of spatial media use. A critical theory-informed philosophy of technology unveils the hegemonic framing of our often taken-for-granted understanding of place. When thinking about human understanding and knowledge of the world and the role of software (such as spatial media) in that understanding in the 'digital age', a computational dimension is inserted into the 'given' (Berry, 2011: 141) or the background of all activities in the world. That 'given' implies that computation is increasingly hegemonic in forming the background for our understanding the world. In essence, software has become 'familiar', something we interact with and are at ease with in our everyday functioning. Our world is increasingly filled with 'actors' enabled with computational techniques (Berry, 2011: 141). These actors may be the desktop PC or stationary computational object, but are increasingly highly mobile computational devices that can be used in navigating the physical world and understanding the physical environment around us.

Screens and interfaces can be understood as a window onto the world, and the keyboard, touchscreen and mouse operate as equipment with which we might manipulate the world. Software and hardware are entwined in our everyday lives, as a result of the trend of the movement of computation 'out of the box and into the environment' (Hayles, 2009: 48). This results in a distributed cognition where applications that perform limited ranges of operations are combined with 'readers' (such as smartphones) that interpret that information, which is linked to databases for storing that information. Spatial media fit within a technological development where internet-enabled mobile computational devices function as part of a wider framework of devices that perform bespoke, discrete tasks in the world and affect subjective understanding and human perception accordingly. Research on this technology typically goes beyond instrumentalist views of technology and looks to understand the changes and remediation of the world through the presence of software in the world.

The revealing of the familiar and yet hegemonic framing of spatial media use focuses attention on the phenomenological understanding of the everyday for users through spatial media. The presence of computation and code in everyday life is manifested through media that allow code to be executed. Coyne describes how digital devices influence the way people use spaces, and argues that such devices are mobile 'tuning' devices in that they draw information into, and out of, situations to help users establish a sense of place (Coyne, 2010: 223). In this way, spatial media 'tune' the user to the place through the incremental changes that the device makes to the experience of place until an 'attunement' to the place is achieved. As such, spatial media and the code that they are dependent upon are part of the user's attunement to the world. Attunement here refers to the feeling of familiarity or being comfortable with our place in the world.

Dodge and Kitchin (2011) offer insights into a different aspect of the relationships between spatial media and their users where the citational, continuously contested and contingent relationships between space, practices and software are highlighted. They draw on the notions of technicity and transduction (Mackenzie, 2002) to argue that the effect of software (or code) on spatial understanding is to continually and contingently modify space; that is, continually bring space into existence through processes of transduction (change) that emerge from the functioning of code and that only imperfectly refresh themselves anew, rendering any spatial relationships only partially and temporarily stable. What this means in practice is that the use of computational devices, as an extension of the rules and instructions of software (code), continually shapes the understanding of place through interactions with coded objects and technologies that bring spaces into existence or awareness. When one opens Google Maps, the space around us or the location we want to explore 'comes forward' into awareness, and is made for us by the functioning of the application. Any omissions in the application will mirror omissions in our understanding of the location if we depend on the application. Code supplements and augments everyday life in this function. The effect of this is to create a series of 'coded practices' that are a combination of code and human practices that become a way of acting and being in the world, specifically in understanding and acting in spaces that are the subject of software and applications.

Many of the issues we discussed in the previous section are concerned with the simultaneous and intersecting flows of data, people and mobile devices, and the consequences of these forms of mobilities to the material practices of enacting and understanding places. This way of theorising the consequence of spatial media draws upon 'the new mobilities turn' that argues that spatial, social, technological and political relations are transformed by multiple, complex and interdependent mobilities (Urry, 2007). This 'mobilities turn' emphasises the co-constitutive relationships between humans and material objects of diverse scales, from urban infrastructure to computer code, drawing on literature in science and technology studies and feminist studies. This literature argues against an all-seeing eye from above and a disembodied cogito (much as the phenomenological approach would) when conceptualising the relationships between humans and material worlds (Haraway, 1991; Latour, 1993). Mobilities studies thus foreground the analysis of the material practices of assembling a wide array of technological objects and humans, and explore how these sociotechnical relationships are differently mobilised. This acts to discern the forces and mechanisms that enable or constrain certain forms of mobilities from the moments, ideas, motivations and temporalities associated with such mobilities. Adopting such a theoretical approach to analyse spatial media use thus pays special attention to how material practices make and remake places according to contingently involved humans and objects. Engaging in location-based games and services, as discussed above, is an example of how place is remade, practices changed and meaning of privacy/publicity contested, because spatial media, humans and contexts of use are contingently and differently assembled. Another example is public participation, and the social and political relationships that it presupposes and draws on, where the participation is performed both in place and across distance when locative media and other mundane information technologies (e.g. internet relay chat or Skype) are appropriated for collaboration between individual participants who do not necessarily share physical proximity for intermittent virtual and face-to-face co-presence to organise tasks and participate in public or artistic interventions.

SPATIAL MEDIA AND FUTURES

As we stand on the precipice of wearable technology, the check-in and manual searching for information using spatial media are becoming passé embodied practices. If we have a location-enabled smartwatch, glasses or clothes, then the recommendations about a place we will receive will be through an interface that provides information instantly, contextually and personally. The harvesting of our social media and computational histories to provide context- and person-specific information in real time will be the mechanism to provide this information. On the one hand, it can be argued that the historical moment of spatial media as mobile devices is coming to a close, and the notion of marking location, sharing location and understanding location as place in this manner will be historic. On the other, social and spatial relationships can be enabled differently or refreshed anew through assembling groups of individuals, sets of technology, and necessary ideas and motivations. Places become dependent upon the collaborative shaping of the movements of spatial media and their users, rapidly growing the data generated from such movements and the multiple (im)mobilities of these data. However, the approach taken, the new technologies (wearables, gesture-based technologies or advanced embodied technologies) will still need to be considered in terms of the subjective, embodied and material experience of place that they co-constitute.

ACKNOWLEDGEMENTS

The research for this chapter and book was funded by a European Research Council Advanced Investigator grant, The Programmable City (ERC-2012-AdG-323636).

REFERENCES

Arthur, C. (2013) 'Apple maps: how Google lost when everyone thought it had won', *Guardian*, 11 November. Available at: http://www.theguardian.com/technology/2013/nov/11/apple-maps-google-iphone-users (accessed 3 January 2015).

Berry, D.M. (2011) *The Philosophy of Software: Code and Mediation in the Digital Age.* London: Palgrave/Macmillan.

Büscher, M., Wood, L. and Perng, S.-Y. (2013) 'Privacy, security, liberty: informing the design of EMIS', in T. Comes, F. Fiedrich, S. Fortier, J. Geldermann and T. Müller (eds), *Proceedings of the 10th International Conference on Information Systems for Crisis Response and Management*, Baden-Baden, Germany. pp. 401–10.

Butcher, M. (2012) 'Welcome to Apple's iOS6 map – where Berlin is now called "Schoeneiche"', *TechCrunch*, 20 September. Available at: http://techcrunch.com/2012/09/20/welcome-to-apples-ios6-map-where-berlin-is-now-called-schoe neiche/ (accessed 26 July 2016).

Campbell, S. and Kwak, N. (2011) 'Mobile communication and civil society: linking patterns and places of use to engagement with others in public', *Human Communication Research*, 37: 207–22.

Coyne, R. (2010) *The Tuning of Place: Social Spaces and Pervasive Digital Media*. London: MIT Press.

Dodge, M. and Kitchin, R. (2011) *Code/Space: Software and Everyday Life*. Cambridge, MA: MIT Press.

Evans, L. (2015) *Locative Social Media: Place in the Digital Age*. London: Palgrave.

Farman, J. (2012) *Mobile Interface Theory: Embodied Space and Locative Media*. New York: Routledge.

Frith, J. (2012) 'Splintered space: hybrid spaces and differential mobility'. *Mobilities*, 7 (1): 131–49.

Fuller, M. (2008) *Software Studies: A Lexicon*. Cambridge, MA: MIT Press.

Galloway, A. and Ward, M. (2006) 'Locative media as socialising and spatialising practices: learning from archaeology', *Leonardo Electronic Almanac*, 14 (3). Available at: http://www.leoalmanac.org/leonardo-electronic-almanac-volume-14-no-3-4-june-july-2006/ (accessed 26 July 2016).

Gordon, E. and de Souza e Silva, A. (2011) *Net Locality: Why Location Matters in a Networked World*. Chichester: Wiley-Blackwell.

Gordon, E., Baldwin-Philippi, J. and Balestra, M. (2013) 'Why we engage: how theories of human behavior contribute to our understanding of civic engagement in a digital era', *Berkman Center Research Publication*, 21: 1–29.

Haraway, D. (1991) *Simians, Cyborgs, and Women: The Reinvention of Nature*. New York and London: Routledge.

Hayles, N.K. (2009) 'RFID: human agency and meaning in information-intensive environments', *Theory, Culture and Society*, 26 (2–3): 47–72.

Hemment, D., Ellis, R. and Wynne, B. (2011) 'Participatory mass observation and citizen science', *Leonardo*, 44 (1): 62–3.

Hjorth, L. (2013) 'Relocating the mobile: a case study of locative media in Seoul, South Korea', *Convergence*, 19: 237–49.

Humphreys, L. (2005) 'Cellphones in public: social interactions in a wireless era', *New Media and Society*, 7 (6): 801–33.

Humphreys, L. (2008) 'Mobile social networks and social practice: a case study of Dodgeball', *Journal of Computer-Mediated Communication*, 13: 341–60.

Humphreys, L. (2010) 'Mobile social networks and urban public space', *New Media and Society*, 12 (5): 763–78.

Latour, B. (1993) *We Have Never Been Modern*. Hemel Hempstead: Harvester Wheatsheaf.

Laurier, E. (2001) 'Why people say where they are during mobile phone calls', *Environment and Planning D: Society and Space*, 19 (4): 485–504.

Licoppe, C. and Inada, Y. (2008) 'Geolocalized technologies, location-aware communities, and personal territories: the Mogi case', *Journal of Urban Technology*, 15 (3): 5–24.

Ling, R. and Haddon, L. (2003) 'Mobile telephony, mobility and the coordination of everyday life', in J.E. Katz (ed.), *Machines That Become Us: The Social Context of Personal Communication Technology*. New Brunswick, NJ: Transaction Publishers. pp. 245–65.

Mackenzie, A. (2002) *Transductions: Bodies and Machines at Speed*. London: Continuum.

McCulloch, M. (2006) 'On the urbanism of locative media', *Places*, 18 (2): 26–9.

Merrin, W. (2014) 'The rise of the gadget and hyperludic media', *Cultural Politics*, 10 (1): 1–20.

Palen, L. and Dourish, P. (2003) 'Unpacking "privacy" for a networked world', in *Proceedings of the SIGCHI Conference on Human Factors in Computing Systems*. New York: ACM. pp. 129–36.

Sheller, M. (2004) 'Mobile publics: beyond the network perspective', *Environment and Planning D: Society and Space*, 22: 39–52.

Shepard, M. (ed.) (2011) *Sentient City: Ubiquitous Computing, Architecture and the Future of Urban Space*. Cambridge, MA: MIT Press.

Silverstone, R. (1994) *Television and Everyday Life*. London: Routledge.

Southern, J. (2012) 'Comobility: how proximity and distance travel together in locative media', *Canadian Journal of Communication*, 37 (1). Available at: http://www.cjc-online.ca/index.php/journal/article/view/2512 (accessed 26 July 2016).

Urry, J. (2007) *Mobilities*. Cambridge: Polity Press.

Williams, R. (2003) *Technology and Cultural Form*. London: Psychology Press.

17

LEVERAGING FINANCE AND PRODUCING CAPITAL

ROB KITCHIN

INTRODUCTION

Ever since maps, gazetteers and almanacs have been created and traded there have been spatialised information economies. With the development of digital data from the 1950s on, the markets for spatial data and information have steadily diversified in products and exploded in volume of trade, with the growth of new market sectors for creating, processing and visualising spatial data such as geographic information systems (GIS) and computer-aided drafting (CAD), and new spatial information products such as geodemographics. This is particularly the case in the Web 2.0 era, with new forms of spatial media such as interactive digital maps, locative social media, city dashboards and augmented reality. Beyond state-produced or subvented forms of spatial media, such as city dashboards or internal or public GIS or open data systems, most spatial media are initiated, developed and owned by private, commercial interests. As such, while many might be deployed as free goods and services at the point of use (e.g. locative social media, some apps, and most websites) they are ultimately concerned with leveraging and producing capital, covering their operating costs in terms of not-for-profit endeavours or turning a profit otherwise. In other words, the spatial media created have to be monetised in some way for continued operation.

Generally, the generation of capital can be produced either directly through the consumer purchase of the spatial media or the services provided via them (for a fixed fee or through a subscription), or indirectly by the sale of data generated from its use or derived data products, or through advertising (with adverts either being pushed to users or being embedded within the product, such as incorporating business details in base maps). Alternatively, the company might derive alternative value through the creation of new products, or through new insights that can be used to improve company efficiencies, productivity or competitiveness. In the case of open data initiatives, such as civic and community endeavours, university repositories or state agency-based initiatives, the product is expected to be free to use, yet the institution is unable to operate like a commercial enterprise, being reliant on state funding or philanthropy or subsidisation from research grants. One way or another, then, developers of spatial media need to find a revenue stream to survive and to justify their investment of time and resources.

It is important then to recognise that finance and capital play an important role in all forms of spatial media: as funding and investment needed to ensure continued operation, and as profit that satisfies investors and shareholders and enables expansion. Moreover, spatial media operate within political economies and regulatory environments, their parameters of operation shaped by government programmes that support start-ups (e.g. state-supported accelerator and incubator programmes) and small and medium-sized enterprises (SMEs) and multinationals (e.g. grants, subsidies, tax incentives and other forms of commercial state aid), tax regimes, laws, licensing and intellectual property regimes. Spatial media thus emerge and are deployed within economic, political and legal contexts. Their rollout also often leads to challenges to those contexts; for example, the generation of vast quantities of indexical geolocational data confronts data protection legislation, and the rise of prosumption (wherein a user acts as both the producer and consumer of the data/service; Ritzer and Jurgenson, 2010) and the sharing economy (where people share or swap or collaborate/co-create resources without being directly employed or formally connected; Botsman and Rogers, 2010) threatens established employment practices and existing business models. This chapter examines some of the economics of spatial media and specifically how they are financed, how they are being used to produce capital, and how they are disrupting existing industries and creating new ones.

SUSTAINING AND DISRUPTIVE INNOVATIONS

Spatial media, in their various forms, constitute a significant economic sector. GIS and digital mapping are large multibillion dollar industries consisting of a diverse ecosystem of companies from the large multinationals such as Esri that provide a range of software and services, through to SMEs that provide more specialist services. All kinds of economic sectors are now deploying spatial media, especially interactive maps, in their websites and apps to engage with customers, provide and source information, and to drive sales. For example, sectors of property sales/rental, and travel, transport and logistics, now rely heavily on online map-driven interfaces to enable customers to discover and explore potential new homes and guest accommodation and possible routes for travel, to book taxis, and to monitor the progress of commercial truck, van and car drivers. *The Economist* (2013) estimates that 3 million jobs in the USA depend on global positioning systems (GPS), a key cornerstone technology for many spatial media, especially those related to satnavs, logistic routers and locative apps on smartphones.

In many cases, the spatial media technologies being produced are sustaining innovations. That is, they provide a more efficient or productive way of performing a task but do not radically transform the work being undertaken. Here, spatial technologies such as GIS are used to produce better knowledge about assets, infrastructure, operations and markets, which is employed to facilitate greater coordination, planning and control within an organisation and to manage it more effectively, efficiently, competitively and productively, while reducing risks, costs and operational losses. Big spatial data gleaned from spatial media are leading to refinements in the products of the geodemographics industry, creating better insights into the spatialised makeup of customers and markets, and helping to refine individual and place profiling and the spatial targeting of goods and services.

In other cases, spatial media are considered disruptive rather than sustaining innovations. That is, rather than maintaining the status quo of how an industry operates or how social relations are configured, they offer a more radical intervention that fundamentally challenges established ways of operating (Christensen, 1997). For example, once GIS became relatively inexpensive it severely disrupted traditional cartography by enabling the creation and querying of bespoke, layered maps in a timely fashion. More recently, the business models of traditional national mapping agencies have been challenged by free-to-access services such as OpenStreetMap, and GIS has been disrupted by online mapping services such as Google Maps that provide free-to-access, interactive mapping. Spatial-media-sharing economy apps such as Uber and Hailo are radically altering how the taxi industry operates within the cities they operate, challenging industry regulation and leading to protests from existing companies and labour unions, who are still employing established technologies and employment practices (White, 2014). In some cases, spatial media, along with related technologies such as big data analytics, have led companies to fundamentally reorganise their structure, but also what the company specialises in. An example is IBM, who in the mid-2000s decided to disinvest from the production of hardware and networked systems and to reorientate its business around analytics and consultancy, with its focus upon what it termed 'Smarter Planet' initiatives.

In other cases, the technology being introduced is not sustaining or disrupting existing industries but is producing new products and services. Augmented reality and locative social media companies had no real equivalent prior to their founding. The spatial media that they produced therefore created an entirely new form of socio-spatial interaction. In the case of Foursquare this quickly scaled to an app that was used by millions of users around the globe (Evans, 2015). In the case of Google and Facebook it is over a billion users, though not all the data generated have a high spatial resolution (i.e. georeferenced with GPS coordinates); however, the metadata include the internet protocol (IP) of the device, which provides some spatial information. But as a free-to-use service, Foursquare's key issue beyond developing their product and building and maintaining a rapidly scaling user base was to devise a means to generate income.

IF THE PRODUCT IS FREE, YOU ARE THE LABOUR AND THE PRODUCT

Different types of spatial media initiatives have different sources and targets for revenue. For digital-mapping companies it is selling a product such as maps or spatial data; for GIS companies it is selling software, consultancy and services; for the emerging sharing economy it is referral fees, the selling of services and monetising the data generated by users; and for locative social media it is advertising revenue and monetising user data. While all new ventures are precarious, the latter are particularly so given the lack of direct funding streams. As such, many spatial media start–ups are reliant on bank loans, angel investors and venture capital, or if they have floated on the stock market, shareholder investment, to stay afloat while they seek a sustainable revenue model. In many cases such a model might not be found. For every Google, Facebook, Airbnb and Uber that develops to become a multibillion dollar company, there are thousands of companies and products that struggle and perish. For example, de Vries et al. (2011) report that the average apps developer makes only US$3000 per year from apps sales, with 80% of paid Android apps being downloaded fewer than 100 times. In addition, they note that even successful apps, such as MyCityWay, which had been downloaded 40 million times, did not generate profits, being sustained by venture capital. Indeed, it may well be that it will take time for new markets to develop and mature. For example, industries underpinned by GPS took many years to blossom after the decision to make the data openly available in 1984.

One of the key means through which many spatial media companies seek to stay afloat is by monetising the personalised data about individuals and companies that they generate. Indeed, the data that the users of their apps divulge – such as their location, personal photos, opinions, ratings, reviews, preferences, values and their network of social contacts – are their key asset, potentially providing a rich insight into their lives, the places they visit and the products they consume. Rather than conduct expensive, sampled consumer surveys where respondents state where they would go and what they would do, such data can be directly harvested from locative social media revealing where all their users actually went and for what purposes, and how they rated the experience (Bollier, 2010). In this sense, a key business model for spatial media is what Zuboff (2015) terms 'surveillance capital-ism' (see Chapters 20, 21 and 22). Moreover, the user of the media acts as a pro-sumer, providing the labour of generating some data at the same time as they are consuming the product. For example, on sites such as tripadvisor.com, prosumers rate and review hotels and other travel services while also consuming these vol-unteered data. The volume of reviews drives additional traffic to the site, generates advertising and referral revenue, and can have a marked influence on the choices of other travellers. It also provides useful data about the individual who volunteered the review, such as their lifestyle choices and travel spending. The insights of such prosumption are of high value to other companies, meaning that the data can be monetised by selling them on to third parties such as data brokers (sometimes called 'data aggregators' 'consolidators' or 'resellers'), who add value by combining them with other data and performing analysis.

Data brokers capture, gather together and repackage data for rent (for one-time use or use under licensing conditions) or resale. By assembling data from a var-iety of sources, including spatial media, data brokers can construct a vast relational data infrastructure that benefits from a 'data amplification' effect (Crampton et al., 2013); that is, data when combined enable far greater insights by revealing asso-ciations, relationships and patterns that remain hidden if the data remain isolated. The size of data holdings of these companies can be huge and is growing rapidly. For example, Epsilon is reputed to own data on 300 million company loyalty card members worldwide (Edwards, 2013). Acxiom is reputed to have constructed a databank concerning 500 million active consumers worldwide, with about 1500 data points per person, and claims to be able to provide a '360-degree view' on consumers (meshing offline, online and mobile data) (Singer, 2012). Other data broker and analysis companies include Alliance Data Systems, eBureau, ChoicePoint, Corelogic, Equifax, Experian, Facebook, ID Analytics, Infogroup, Innovis, Intelius, Recorded Future, Seisint and TransUnion.

Each company tends to specialise in different types of data and data products and services. Products include lists of potential customers/clients who meet certain criteria and consumer and place profiles; search and background checks; derived data products wherein brokers have added value through integration and analytics. In addition data analysis products include the micro-target advertising and marketing campaigns (by social characteristics and/or by location) that guide the building of long-term customer relationships through personalised experiences, assess credit worthiness, socially and spatially sort individuals (e.g. shaping whether an individual is cultivated as a customer through incentives or whether a person gets a loan, or a tenancy or a job), and provide tracing services, and supply detailed business analytics. Further, companies seek to predictive model what individuals might do under different circumstances and in different places, or calculate how much risk a person constitutes (CIPPIC, 2006; see Chapter 21). Moreover, such data can also be used to set the parameters for dynamic pricing and how much a person might expect to pay for goods and services, given their profile.

The worry of some, including Edith Ramirez (2013), the chairperson of the Federal Trade Commission (FTC) in the USA, is that such firms practise a form of 'data determinism' in which individuals are not just profiled and judged on the basis of what they have done and where they live, but also on the prediction of what they might do in the future. In other words, there is concern that the data that spatial media and other technologies produce will precede their users, having all kinds of implications with regards to how they are treated by companies and the state (see Chapters 20, 21 and 22). Interestingly, given the volumes and diversity of personal and place-based data that spatial media produce, and that data brokers and analysis companies possess, and also how the data are used to socially and spatially sort and target individuals and households, there has been remarkably little critical attention paid to their operations. Indeed, there is a dearth of academic and media analysis about such companies and the implications of their work and products.

FINANCING OPEN ACCESS SPATIAL MEDIA

As noted, many spatial media do not directly charge consumers (e.g. locative social media) and not all spatial media are commercial in orientation (e.g. many online GIS or city dashboards are produced by state agencies and universities; many public-service apps are state sponsored; and initiatives such as OpenStreetMap are voluntary, community ventures). The challenge in these cases is to find an alternative means to finance the enterprises in the absence of directly charging for use. In the case of commercial enterprises, as discussed, the route pursued is usually to monetise the data and to sell additional services. The challenge for open access spatial media is more onerous as they adopt a different ethos that prioritises the public good (see Chapter 9). Open access in its purest form is 'digital, online, free of charge, and free of most copyright and licensing restrictions' (Suber, 2013). In other words, it seeks to remove both 'price barriers (subscriptions, licensing fees, pay-per-view fees) and permission barriers (most copyright and licensing restrictions)' (Suber, 2013) so that the spatial media and their associated data are freely available 'on the public internet' and can be used for 'any lawful purpose, without financial, legal, or technical barriers other than those inseparable from gaining access to the internet itself' (Budapest Open Access Initiative, 2002). As such, the data as well as the media are open access and cannot be monetised.

The conundrum for open access spatial media then is to find a way to deliver the technology and services with no or limited for-fee income. The solution has been the creation of a variety of open access positions that take varying stances on issues such as permission barriers, timing, who pays and how for production, and the use of a variety of funding models. Different open access models include gratis open access (free of charge, but not free of copyright of licensing restrictions), libre open access (free of charge and expressly permitting uses beyond fair use), delayed open access (paid access initially, becoming open after a set time period), green and gold open access (pay-for-production followed delayed publication in an open access repository or gratis open access) and so on (Suber, 2013). Kitchin et al. (2015) identify 14 different potential funding sources for funding open access endeavours, which they group into six classes (institutional, philanthropy, research, audience, service and volunteer; see Table 17.1).

Different types of open access projects use different combinations of these funding sources. For example, initiatives such as OpenStreetMap rely extensively on volunteered labour and philanthropic donations to fund their work. In contrast, a university initiative such as AIRO (All-Island Research Observatory; which provides online spatial data visualisation of Irish public administration datasets) relies on core funding, research grants, consultancy and white-label development. The issue for all open access initiatives is that, beyond core state funding, the finance streams they rely on are cyclical and uncertain, and while often very worthy they find it difficult to raise and maintain continual funding, placing them under stress and threatening their existence. There is therefore a real question as to the sustainability of many open access spatial media initiatives and it will be interesting to observe how many will be continuing to operate in 10–15 years' time.

ISSUES FOR FURTHER REFLECTION

To date, much of the research and thinking about spatial media has focused on understanding the media themselves, their uses, their effects on individual spatial behaviour and knowledge, and how space and socio-spatial relations are surveilled and governed. There has been very little focus on the economic geographies of spatial media and how spatial media challenge existing social, political and economic contexts or established business models. Different forms of spatial media

TABLE 17.1   Models of funding open access spatial media

	Model	Description
Institutional	Core funded	The state provides the core operational costs through a subvention as with other state data services such as libraries, national archives, statistical agencies, etc.
	Consortia (membership) model	Build a consortium that collectively owns the data, pools labour, resources and tools, and facilitates capacity building, but charges a membership fee to consortium members to cover shared value-added services
	Built-in costs at source	When research grants are awarded by funders, applicants must build in the costs for archiving the data and associated outputs in a repository at the end of the project. This funding is transferred to the repository for any services rendered
	Public/private partnership	Public/private partnerships, with the public sector providing the data and private companies providing finance and value-added services for access and re-use rights
Philanthropy	Philanthropy/ corporate sponsorship	Funding is sourced from philanthropic organisations as grants, donations, endowments and/or corporate sponsorship. If an endowment is sizable then core services can be funded from the interest. The donations can also be used to leverage other funding, for example, matched money from the state. This can also be reversed, so that state funding is used to try to leverage philanthropic funding/corporate sponsorship
Research	Research funded	The majority of funding is generated through the sourcing of research grants from national and international sources, with overheads being used to subvent core services
Audience	Premium product/ service	Offers end-users a high-end product or a service that adds value to data (e.g. derived data, tools or analysis) for payment, either as fixed payment, recurrent fees or pay-per-use, without using monopoly rights. This enables the data producer to gain first-mover advantages in the marketing and the sale of complementary goods
	Freemium product/ service	Offers end-users a graded set of options, including a free-of-charge option that includes basic elements (e.g. limited features or sampled dataset), with more advanced, value-adding options (e.g. special formats, additional functionality, tools) being charged a fee. Opens up the product/service to a wider, low-end market and more casual use, while retaining paid, high-end product/service for more specialised users
	Content licensing	Make the data free for non-commercial re-use, but charge for-profit re-users
	Infrastructural razor and blades	An initial inexpensive or free trial is offered for products/services (razor), which encourages take-up and continued paid use (blades). It might be that access is free through application programming interfaces (APIs), but that computational usage is charged on a pay-as-you-go model, with the latter cross-subsidising the former

	Model	Description
Service	Pay per purpose	Charge for services beyond data use, such as ingest, archiving, consulting and training services
	Free with advertising	Products/services are provided for free, but users receive advertising when using the product/service (revenue generating) or the products/services are provided by different companies and branded as such to encourage use of their other products/services (cross-subsidisation)
	White-label development/ platform licensing	A customised product/service is created for a client and branded for their use, with that client paying a one-off fee or subscription that includes maintenance and update costs
Volunteer	Open source	Offers end-users data products/services for free, with the infrastructure maintained on a voluntary basis, including crowdsourcing

Source: Kitchin et al. (2015). Assembled from Ferro and Osella (2013); Maron (2014); consultation with Digital Repository of Ireland stakeholders and team discussion.

operate as both sustaining and disruptive innovations and are deeply implicated in the emergence of the new sharing economy, as well as economic sectors such as data brokerage, security and surveillance. As yet, however, we have little empirical understanding or detailed theoretical knowledge of the ways in which spatial media create and sustain disruption in relation to business models, company organisation and operation, labour practices, markets, consumption and regulatory oversight. Nor do we know how these disruptions might vary between different forms of spatial media, or how they are unfolding in different places as a function of varying culture, governance, legislation and political economies. Moreover, we know little about the geography of spatial media production – for example, are there agglomerations of spatial media industries, and are they similar to related industries such as software production more generally? Are there global divisions of labour in the production of spatial media? In short, there is a real need to conduct economic geographies of spatial media.

Further, as discussed in Chapters 20, 21 and 22, there is also a need to examine in more detail the social and political implications of the economics associated with spatial media. While the sharing economy is often lauded as a new gift economy that promotes collaboration and community building, it has also been critiqued as being co-opted by market capitalism as a way of accessing volunteered labour, and casualising labour, and circumventing some of the costs of doing business such as avoiding taxes, regulations and insurance. Data brokers gather up huge quantities of spatial media data, conjoin them with other data, and apply analytics that profile, target, and socially sort individuals with and places. They are little regulated and are not required by law to provide individuals with access to the data held about them, nor are they obliged to correct errors relating to those individuals. The data industry then raises all kinds of ethical questions about privacy, how its work affects people, and how it might be regulated in an age where data production is ubiquitous. The move to create open access spatial media, either as state endeavours or as volunteered, crowdsourced initiatives, is not straightforward. While the spatial media represent a public good, there are still associated costs with their production that need to be financed and there are knock-on consequences in making data open for established data providers. As such, there is a more complicated set of political, social and economic implications associated with open access than many of its proponents would acknowledge (see Kitchin, 2014). In short, a whole series of studies is required concerning the implications of the new economies produced by spatial media and how regulations should be altered to minimise their more pernicious effects.

ACKNOWLEDGEMENTS

The research for this chapter and book was funded by a European Research Council Advanced Investigator grant, The Programmable City (ERC-2012-AdG-323636).

REFERENCES

Bollier, D. (2010) *The Promise and Peril of Big Data*. Aspen, CO: The Aspen Institute. Available at: http://www.aspeninstitute.org/sites/default/files/content/docs/pubs/The_Promise_and_Peril_of_Big_Data.pdf (accessed 27 July 2016).

Botsman, R. and Rogers, R. (2010) *What's Mine Is Yours: The Rise of Collaborative Consumption*. New York: Harper Business.

Budapest Open Access Initiative (2002) 'Read the Budapest Open Access Initiative'. Available at: http://www.budapestopenaccessinitiative.org/read (accessed 27 July 2016).

Christensen, C.M. (1997) *The Innovator's Dilemma*. Cambridge, MA: Harvard Business Review Press.

CIPPIC (2006) 'On the data trail: how detailed information about you gets into the hands of organizations with whom you have no relationship. A report on the Canadian Data Brokerage Industry'. Ottawa: The Canadian Internet Policy and Public Interest Clinic. Available at: https://cippic.ca/sites/default/files/May1-06/DatabrokerReport.pdf (accessed 11 August 2016).

Crampton, J., Graham, M., Poorthuis, A., Shelton, T., Stephens, M., Wilson, M.W. and Zook, M. (2013) 'Beyond the geotag? Deconstructing "Big Data" and leveraging the potential of the geoweb', *Cartography and Geographic Information Science*, 40 (2): 130–9.

de Vries, M., Kapff, L., Negreiro Achiaga, M., Wauters, P., Osimo, D., Foley, P., Szkuta, K., O'Connor, J. and Whitehouse, D. (2011) 'Pricing of Public Sector Information Study (POPSIS)'. Available at: http://ec.europa.eu/newsroom/dae/document.cfm?doc_id=1158 (accessed 11 August 2016).

The Economist (2013) 'Open data: a new goldmine', *The Economist*, 18 May. Available at: http://www.economist.com/news/business/21578084-making-official-data-public-could-spur-lots-innovation-new-goldmine (accessed 27 July 2016).

Edwards, J. (2013) 'Facebook is about to launch a huge play in "Big Data" analytics', *Business Insider*, 10 May. Available at: http://www.businessinsider.com/facebook-is-about-to-launch-a-huge-play-in-big-data-analytics-2013-5 (accessed 27 July 2016).

Evans, L. (2015) *Locative Social Media: Place in the Digital Age*. London: Palgrave.

Ferro, E. and Osella, M. (2013) 'Eight business model archetypes for PSI re-use. Open data on the web', Workshop, 23–24 April 2013, Google Campus, Shoreditch, London. Available at: http://www.w3.org/2013/04/odw/odw13_submission_27.pdf (accessed 27 July 2016).

Kitchin, R. (2014) *The Data Revolution: Big Data, Open Data, Data Infrastructures and their Consequences*. London: Sage.

Kitchin, R., Collins, S. and Frost, D. (2015) 'Funding models for open access digital data repositories', *Online Information Review*, 39 (5): 664–81.

Maron, N. (2014) 'A guide to the best revenue models and funding sources for your digital resources', Ithaka S+C and JISC. Available at: https://www.jisc.ac.uk/reports/the-best-revenue-models-and-funding-sources-for-your-digital-resources (accessed 27 July 2016).

Rameriz, E. (2013) 'The privacy challenges of big data: a view from the lifeguard's chair', Technology Policy Institute Aspen Forum, 19 August. Available at: https://www.ftc.gov/sites/default/files/documents/public_statements/privacy-challenges-big-data-view-lifeguard%E2%80%99s-chair/130819bigdataaspen.pdf (accessed 27 July 2016).

Ritzer, G. and Jurgenson, N. (2010) 'Production, consumption, prosumption: the nature of capitalism in the age of the digital "prosumer"', *Journal of Consumer Culture*, 10 (1): 13–36.

Singer, N. (2012) 'You for sale: mapping, and sharing, the consumer genome', *New York Times*, 17 June. Available at: http://www.nytimes.com/2012/06/17/technology/acxiom-the-quiet-giant-of-consumer-database-marketing.html (accessed 27 July 2016).

Suber, P. (2013) 'Open access overview'. Available at: http://legacy.earlham.edu/~peters/fos/overview.htm (accessed 27 July 2016).

White, J. (2014) 'Hailo, Uber and the deregulation of Ireland's taxi industry', Programmable City blog, 11 June. Available at: http://www.maynoothuniversity.ie/progcity/2014/06/hailo-uber-and-ireland/ (accessed 27 July 2016).

Zuboff, S. (2015) 'Big other: surveillance capitalism and the prospects of an information civilization', *Journal of Information Technology*, 30: 75–89.

18

OPENNESS, TRANSPARENCY, PARTICIPATION

TRACEY P. LAURIAULT AND MARY FRANCOLI

INTRODUCTION

In this chapter openness, transparency and participation are conceptually framed within the discursive regime of open government. Spatial media are understood as an evolving concept which include: geographic information systems (GIS) and maps; as media (Peterson, 1995; Sui and Goodchild, 2001; Wilson and Stephens, 2015), the spatial mediation of heterogeneous content (Cartwright et al., 2007); a way to 'mediate seeing' at a distance (Fremlin and Robinson, 1998); and cartographic mediation and the processes of geomediation (Pulsifer, 2005). In the open government space, spatial media are used to communicate how agencies are performing with respect to key indicators, to illustrate the spatial arrangement of policy adopters and violators, and to benchmark how nations fare relative to each other on issues related to transparency, accountability and citizen engagement. Spatial media also include interfaces used to navigate institutional webpages and wade through data, information and knowledge (Elwood and Leszczynski, 2012). In addition, spatial media are mobilised by actors to laud their progress, which inadvertently (or perhaps advertently) creates competition between agencies and states. They are also seen as dynamic global maps that logo-ise (Anderson, 1991) the reach of transnational institutions and brand the power/knowledge of openness, transparency and participation indicator systems.

Spatial media are used to support a mild technocratic ideology, embraced by national governments, and to some extent transnational organisations such as the G8 and the World Bank, but are also used to support progressive technological politics (Feenberg, 2011). Civil society organisations using spatial media have embraced new managerialism techniques to monitor progress and accountability (Craglia et al., 2004), or as evidence to support their own mandates. However, in doing so, they can adopt the mechanically objective stance of public-sector administrators (Porter, 1995) and may be masking complex, nuanced and contextual information politics (Kitchin et al., 2015). Spatial media are therefore enlisted to simplify access to the data and indicators of this 'progressive' and 'democratic' process, thus perpetuating the established practice of celebrating maps as unbiased truth and fact (Harley, 1989; Crampton, 2001).

This chapter proceeds by first defining openness, transparency and engagement. A selection of spatial media examples are then used to inform the discussion, concluding with some critical reflections on the nature of this form of spatial communication.

OPENNESS, TRANSPARENCY AND PARTICIPATION

OPENNESS

The notion that governments should be 'open' and disclose information and data to citizens is not new. Legislation granting the public access to government documents existed as early as 1766 in Sweden. In the USA, Wallace Parks espoused the importance of open government in the 1950s during the drafting of the Freedom of Information (FOI) legislation enacted in 1966. Since then national and subnational governments in at least 100 countries[1] have adopted their own freedom of information legislation or guidelines. In addition, institutions have been established in the form of information commissioners or ombuds in many jurisdictions to provide oversight of access regimes and to act as advocates for those citizens, journalists or civil society organisations who believe they have been wrongfully denied access to information, or who are having trouble navigating their way through what appears at times to be an overly complicated and bureaucratic informational system. FOI was seen as key to opening government and making it more transparent. In the USA, FOI was aggressively pursued by Congressman John Moss, who claimed that widespread government secrecy was leading to problems in the USA during the Cold War, while in the UK, FOI was primarily geared towards access to unstructured information such as public records, government communications and expenditures.

FOI is thought to lead to the improvement of democracy. While there is an association between greater openness and democracy, this may be an overly deterministic assumption and, at least empirically, causality may be overstated (Garsten and Lindh de Montoya, 2008). Classic contemporary examples are Edward Snowden and Julian Assange, who come from democratic states but are facing prosecution as a result of sharing information and data they considered to be in the public interest to know, but which the state was concealing. Their actions have had undemocratic consequences, whereas less democratic states with open government and open data strategies, such as Russia,[2] are applauded (however, this might be a form of 'information' politics as Russian journalists can attest that the amplification of data and information used as evidence of infractions, preferential treatment or corruption has led to very dire consequences). Concurrently, being able to decipher and understand the mechanics and instruments of open government requires significant levels of legal, technical and administrative expertise that elude many citizens (Straumann, 2014; Brownlee and Walby, 2015).

While FOI regimes were largely established after Parks' (1957) writing, the term 'open government' itself seemed to fall out of favour until 2009 when it was employed by US President Barack Obama when issuing an Open Government Directive[3] to make government information open and accessible by default. The term gained further momentum in 2011 with the establishment of the Open Government Partnership (OGP),[4] an international initiative that seeks commitments from countries around the world to improve openness.

With the establishment of the OGP, FOI legislation is no longer the defining feature of open government as it was for Parks. Among other things, the OGP also champions the importance of citizen participation, as well as open data (see Chapter 9). Today, there are 70 OGP member countries that have endorsed the *Open Government Declaration*, which carefully points to open information, open data and citizen engagement. In signing the Declaration, countries commit to:

- 'promoting increased access to information and disclosure about governmental activities at every level of government;

- increasing efforts to systematically collect and publish data on government spending and performance for essential public services and activities;

- pro-actively providing high-value information, including raw data, in a timely manner, in formats that the public can easily locate, understand and use, and in formats that facilitate reuse;

- providing access to effective remedies when information or the corresponding records are improperly withheld, including through effective oversight of the recourse process;

- recognizing the importance of open standards to promote civil society access to public data, as well as to facilitate the interoperability of government information systems;

- seeking feedback from the public to identify the information of greatest value to them, and pledge to take such feedback into account to the maximum extent possible.'[5]

The rapid increase in government and civil society participation in the OGP signals its growing importance in shaping the debate and actions around issues of openness. However, it is worth noting that openness is not necessarily about evidence-informed public engagement in formalised processes such as environmental and social impact assessments (ESIA), or about creating evidence-informed open dialogue between citizens and governments more broadly on issues beyond open data or open government. Also notable, the OGP is a form of progressive technological politics, but what it measures and how it measures can potentially mask subtle forms of regressive politics. Canada, for example, has an illustrious reputation when it comes to open government and open data, even though the Harper government under which these programmes flourished was renowned for silencing public servants, quashing science, impeding access to information and shutting down data collection institutions, such as the mandatory long-form census and environmental monitoring stations. This was problematic to the point that evidence-informed decision-making and open government became political platforms for political parties in the 2015 elections (O'Hara and Dufour, 2014; Lauriault and O'Hara, 2015) and social movements such as *Evidence for Democracy*.[6] Members of the OGP Independent Review Mechanism have also advocated for changes in how they evaluate national programmes (Francoli, 2014).

It is also important to note that OGP membership is currently restricted to national governments.[7] For unitary states, national membership does not come with the same subjurisdictional issues as seen in federations such as Canada, Russia, Australia and many others, where there are distinct divisions of power between different levels of government.[8] National governments have a mix of formal engagement mechanisms, beyond consultations such as House and Senate committees and subcommittees, commissions, inquiries and, as mentioned, SEIA.[9] However, consultation within the OGP framework is often limited, with governments simply informing and not fully involving citizens (Francoli et al., 2015).

In terms of spatial media and the OGP, users can click the 'Countries' tab on the OGP home page to get to an interactive choropleth map of the world (Figure 18.1), depicting where members are in terms of the 1st, 2nd and 3rd action plan submission cycle and which countries are developing plans. Below this map is a full list of member countries. Clicking on a country name navigates the user to a country page that includes: an overview of where they stand, access to records such as action plans, self-assessment reports, progress reports, case studies, contact information and a checklist of accomplishments.[10]

The OGP provides easy access to the data generated by its independent reporting mechanism, which assesses national action plans and their completion via the 'OGP Explorer'. The Explorer builds on knowledge produced by developers from key civil society actors such as the World Wide Web Foundation, and it is funded by the International Development Research Council (IDRC), both of which produce their own indicators and benchmarks related to open data.

The OGP exemplifies the use of spatial media as an interface to the institution and to its treasure trove of data, indicators and information; as means to logo-ise openness, transparency and participation; and as a way to readily visualise technological politics. Because of their graphical use in this context, spatial media do not reveal the subtle geopolitics of this information space, however. This more nuanced information can be found in other narratives on the OGP site, particularly in its independent country reports.

It is important to note that while the OGP is a fairly new initiative there are others that have been working towards open government for a longer time without explicitly using the term 'open government'. *Open access* is a prime example. It is a 'worldwide effort to provide free online access to scientific and scholarly research literature, especially peer-reviewed journal articles and their preprints'.[11] According to the *Open Access Directory* timeline[12] it started in 1966, and lists hundreds of activities, events, journals and legislation related to open access. Although open access differs from the discursive regime of open government, national research-funding policies that mandate access to the results of research such as data, and promote data

Participating Countries

OGP was launched in 2011 to provide an international platform for domestic reformers committed to making their governments more open, accountable, and responsive to citizens. Since then, OGP has grown from 8 countries to the 69 participating countries indicated on the map below. In all of these countries, government and civil society are working together to develop and implement ambitious open government reforms.

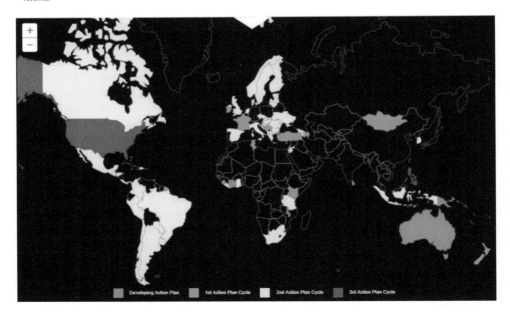

FIGURE 18.1   OGP map of participating countries

management plans, preservation and the publishing of research in open access journals, are pragmatically framing these under their national open government banners. Additionally, there is the *Open Society Foundation*, which promotes many of the ideals linked to open government. The concept of an *open society* is somewhat broader than open government. In its simplest form it stands for 'human rights and balanced values that include autonomy, accountability, privacy and … responsible secrecy' (Holzner and Holzner, 2006: 3).

TRANSPARENCY

Transparency may be understood as 'the social value of open, public, and/or individual access to information held and disclosed by centers of authority' (Holzner and Holzner, 2006: 33). It is associated with accountability, with a citizen's right to know what the government is doing, and with the duty of government to disclose information. It is grounded in notions of trust. While openness and transparency are closely related concepts, policies, practices and actors differ. FOI, for example, is a form of government transparency. The practice of proactive disclosure, however, is more closely associated with the articulations of transparency; this includes the publishing of government budgets, spending and contracting, which is thought to mitigate corruption. Other examples are the proactive disclosure of environmental hazards, contaminated sites and waste depots.

Like openness, transparency has long historical roots. It is associated with the enlightenment, the scientific revolution and scholarly inquiry, and also literacy, freedom of speech, freedom of expression, freedom of the press, and changes in authoritative and legal structures, as well as with the rise of democracy and the need for public knowledge, such as opening the census or population health. In the 20th century, transparency was normalised with applied social science research, programme evaluation and outcomes-based management techniques. This was the era where measures and techniques were developed to assess how government was performing, often for government and by government, but nonetheless enabling the assessment of trust at a distance, and arguably also a form of mechanical objectivity. Administrations and their programmes needed to be accountable to the electorate (Lord, 2006). The enablers of transparency are considered to be: the spread of demo-cratic governance; the rise of global media; the growth and spread of non-governmental organisations; the proliferation of regimes requiring governments to disclose infor-mation; and the widespread use of information technologies (Lord, 2006).

Transparency is most often framed as shining a light on ignorance and corruption, as well as the promotion of trust and accountability. It comes with challenges, such as privacy (see Chapter 22), the possibility for greater surveillance, and the creation of a chilly climate where public administrators and politicians, for fear of FOI and possible reprisals, choose not to take meeting minutes and not to communicate on platforms over which the government has jurisdiction, which has led to the empty-archives syndrome and overly cautious administrators (Worthy and Hazel, 2014).

There has been a rise in organisations advocating for, and working within the framework of, transparency. Perhaps most notable internationally is Transparency International (TI), a global non-profit organisation involved in the anti-corruption movement. It has enabled a global cultural shift towards greater transparency and accountability. Though established much earlier than the OGP, TI is building on the momentum of the OGP to establish its own open government project to help citizens understand and participate in governance. It advocates for the national and inter-national adoption of open governance standards, and monitors compliance with these standards. TI produces a number of transparency indices and scorecards and enlists spatial media to amplify its results. The Corruption Index map as seen in Figure 18.2 is an example of this. It illustrates the degree of corruption along a colour spectrum, with red representing highly corrupt states and yellow for highly clean states. TI encour-ages users to broadcast this spatial media by providing basic code[13] to facilitate the embedding of the map into other social media communication platforms such as blog posts, or the re-use of the map's underlying data in apps via the provision of data in JavaScript Object Notation (JSON) format.[14] The complete Corruption Index data can be downloaded as an Excel spreadsheet, with countries assessed along a number of internationally used and accepted indices and indicators. In addition, TI produces a more complex interactive map as an interface to more detailed country analysis, seen in Figure 18.3a. Once selected, the map zooms into a crude graphical rendering

about people's lives. And as the map below shows, it's a global problem.

FIGURE 18.2   Transparency International embeddable global logo map

FIGURE 18.3A   Transparency International interactive corruption by country/territory interface map

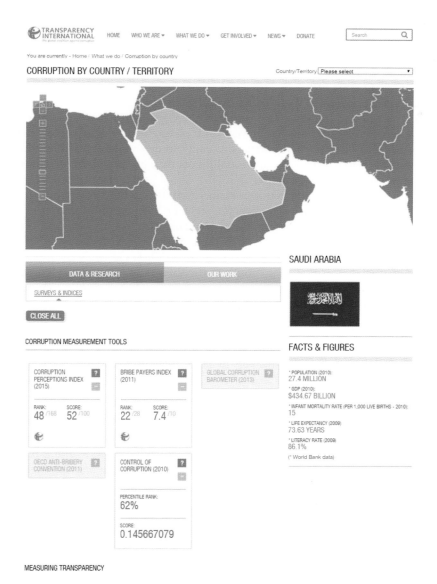

FIGURE 18.3B Saudi Arabia – Transparency International interactive corruption by country/territory

of a country's political boundaries and below it key indicators are featured in a readily accessible and interactive infographics; such indicators include its own *Corruption Perception Index*, a *Bribe Payers Index*, adherence to the *OECD Anti-Bribery Convention*, the Natural Resource Governance Institute (NRGI), Brookings Institution and World Bank Development Research Group *Control of Corruption Index*, the *Global Corruption Barometer* public opinion survey index, and many others (Figure 18.3b).

The TI webpage is an information gateway to numerous internationally recognised indices and indicators. It provides access to its methodologies, reports and data, and disseminates these in a way that non-experts can re-use. These spatial media can be mobilised as evidence by transparency advocates in all sectors, and often reaffirm what citizens in many countries already know. They are very powerful instruments used by transnational organisations to shape international investment and overseas development aid and audit, and to monitor the operational practices of businesses in 'corruption clean' states in 'highly corrupt' places. TI has managed to make the complex and dirty practices of corruption easily understandable by rendering these into easy-to-comprehend and clean spatial media utilising popular notions of online aesthetics and navigation techniques.

Access Info Europe is another similar organisation. It is a 'human rights organisation dedicated to promoting and protecting the right of access to information in Europe as a tool for defending civil liberties and human rights, for facilitating public participation in decision making, and for holding governments accountable'.[15] *Access Info Europe* has been coordinating the development of an *Open Government* standard[16] with other civil society actors that are centred on the themes of transparency, participation and accountability. In a recent publication, its Executive Director critically examined the role of transparency indicators and points out that, even though access to information laws have been in place since the 18th century, the United Nations has only just recently linked these to freedom of expression, and suggests that the most significant advances have been spearheaded by leaks and whistleblowers such Snowden and WikiLeaks or by code-breaking hackers (McGonagle and Donders, 2015).

PARTICIPATION

Democracy is premised on the notion of active engagement between citizens, public administrators and elected government officials. Arguably, openness and transparency as framed in this chapter are understood to be ways to strengthen and bridge the gap between citizen and state in terms of access to actionable knowledge that enables a reasoned and informed dialogue with government. Openness and transparency can also be characterised as the legal, technological and normative framework that allows for the public to participate in policy-making, which includes access to data, information, a clear methodology and a well-facilitated engagement process.

Meaningful and deep forms of engagement are more than a one-way dialogue. Meaningful participation includes the engagement of all stakeholders who openly and transparently share their knowledge resources to ensure the best possible course of action, leading to mutual gain and the social, economic and environmental wellbeing of all those concerned (Sheedy, 2008). From a technocratic perspective, participation would be considered as a slow and inefficient way to govern. From a new managerialism perspective it might be considered as unpredictable and unwieldy. And to a lesser extent, those advocating for a more objective approach to governing might consider participation a way to infuse subjective politics into the equation. From a technological politics perspective, as advocated by Andrew Feenberg (2011), meaningful engagement is a way to democratically mediate interests for the public good and to consider public interests beyond engineered notions of progress and efficiency.

There has been an effort among policy-makers, and various civil society organisations, including many involved with the OGP, to improve engagement mechanisms and frameworks. *Open by default* is an example. It is a socio-technological framework to design engagement as part of open government strategies (Government of Ontario, 2014). It builds on the principles of *privacy by design* (Cavoukian, 2011) embraced by privacy commissioners and data protection officers, and *access by design* (Cavoukian, 2010) advocated by information commissioners and open data enthusiasts. Open by default implies the creation of an open and transparent engagement infrastructure within government, where deeper levels of engagement may be technologically mediated and where government as an institution is open to public participation and participatory evidence-informed policy-making. How this will be spatially mediated and what platforms will be used are uncertain, although some early examples can be found in the form of environmental and social impact assessments (ESIAs).

ESIAs are an established, formalised and deep form of open, transparent and heavily regulated public participation and engagement. Here, GIS and spatial media

FIGURE 18.4a   United Nations e-participation interactive global logo interface map

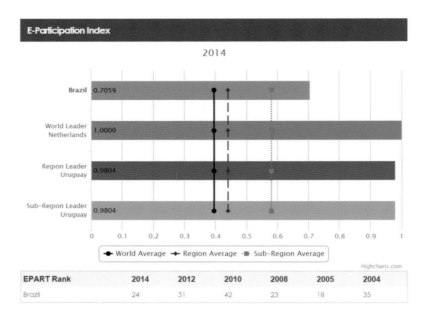

FIGURE 18.4B   UN e-participation index ranking chart from the dedicated country page of Brazil

Source (for a & b): 'United Nations E-Government Survey 2014: E-Government for the Future We Want' by Department of Economic and Social Affairs, © 2014 United Nations. Reprinted with the permission of the United Nations.

are used as boundary objects (Star and Griesemer, 1989) to facilitate understanding and to geographically focus dialogue about the space, place and location where a large infrastructure project is proposed. ESIAs aim to predict the possible outcomes of programmes, projects and policies in order to mitigate and reduce negative impacts and/or identify and amplify positive outcomes. These are procedurally complex processes demanding a deep commitment by all those involved. They also necessitate scientific, technical, professional and local expertise, all of which are data and information intensive. These processes can be said to be open by default as they allow for all stakeholders to have informed and rational dialogue based on the same facts and information. The ideal is, however, much more complicated, and there are significant power imbalances in terms of knowledge, human resources and finances between government and private-sector actors supporting a particular project, and the local population who may be its recipients. Other spatially mediated forms of open and transparent participation processes are citizen science, Volunteered Geographic Information (VGI) and participatory mapping (see Chapters 12 and 13).

Many different mechanisms have been developed to measure engagement. One of the longer-standing comparative examples is the United Nations Public Administration Network's (UNPAN) E-Participation Index (see Figure 18.4a). UNPAN has developed a number of e-government indices. This one measures participation from passive to deep forms of engagement as follows: e-information assesses the provision of public information and access to information upon demand; e-consultation assesses contributions to and deliberation on public policies and services; and e-decision-making assesses empowerment through co-design of policy options and co-production of service components and delivery modalities.[17] The dynamic map interface mediates access to a dedicated country page (Figure 18.4b), with an inset map, demographics and a series of charts that compare how that country fares on an index relative to the highest ranking in the world, the regional leader and the subregional leader.

The International Association offers another tool for evaluating or categorising engagement for public participation, or IAP2.[18] Its mandate is to promote improved participation across a range of institutions that impact the public. The IAP2 has developed a spectrum outlining varying levels of engagement. As can be seen in Table 18.1, these include: 'inform', 'consult', 'involve', 'collaborate' and 'empower'.[19] Inform, at one end of the spectrum, includes no significant engagement. Rather, the public is simply provided with information. The opportunity for engagement

TABLE 18.1 IAP2 public participation spectrum[19]

	Inform	Consult	Involve	Collaborate	Empower
Public participation goal	Provide the public with balanced and objective information to assist them in understanding the problems, alternatives and/or solutions	Obtain public feedback on analysis, alternatives and/or decision	Work directly with the public throughout the process to ensure that public concerns and aspirations are consistently understood and considered	Partner with the public in each aspect of the decision including the development of alternatives and the identification of the preferred solution	Place final decision-making in the hands of the public
Promise to the public	Keep you informed	Keep you informed, listen to and acknowledge concerns and aspirations, and provide, feedback on how public input influenced the decision	Work with you to ensure that your concerns and aspirations are directly reflected in the alternatives developed, and provide feedback on how public input influenced the decision	Look to you for advice and innovation in formulating solutions and incorporate your advice and recommendations into the decisions to the maximum extent possible	Implement what you decide

increases marginally moving through the spectrum, with the highest level of engagement being 'empower'. Here, citizens are provided with the ability to make decisions that will then be implemented. The IAP2's spectrum has become an international standard for engagement and is increasingly being used within OGP discussions and research (Francoli et al., 2015).

CONCLUSION

One of the major challenges for those researching and working in the areas of openness, transparency and engagement has been to get a broad overview of the state of these issues in various jurisdictions and to establish mechanisms for monitoring and measuring progress. In this chapter, we defined these concepts in the context of spatial media. We provided examples and showcased the mediated seeing, interaction with and navigation through data, indicators and indices from the OGP, TI and UNPAN. A number of examples of mediated engagement were discussed, and we provided visual depictions of the following concepts: maps as media, the spatial branding of power/knowledge, and the map as interface. Further critical evaluation of the veracity, validity and quality of the openness, transparency and participation data and indicators that were spatially mediated is discussed by Kitchin et al. (2016).

Spatial media in the context of open government are used for evidence-informed decision-making at a high level, but not necessarily to inform a critical analysis of the underlying subnational contexts of administrative, political and knowledge-producing cultures. Here, we discussed how spatial media advance the soft technocratic ideologies and new managerialism epistemologies, often considered as inherent in this open government space; but, on a more optimistic note, we demonstrated too that this techno-social open government space can also be one in which critical and progressive technological politics can be spatially mediated, seen, advanced and re-mediated.

NOTES

1. FreedomInfo.org lists 105 FOI regimes between 1766 and 2015 (http://www.freedominfo.org/). Theirs is not the only count, and numbers differ depending on the nature of definitions, as discussed in one of their earlier 2012 reports – http://www.freedominfo.org/2012/10/93-countries-have-foi-regimes-most-tallies-agree/
2. See World Bank Support – http://www.worldbank.org/en/news/feature/2015/01/13/unleashing-the-potential-of-open-data, http://data.gov.ru/, and the GosZatraty Clear spending project – http://clearspending.ru/page/about/en/
3. http://www.state.gov/open/
4. http://www.opengovpartnership.org/
5. The full Declaration can be found at: http://www.opengovpartnership.org/about/open-government-declaration
6. https://evidencefordemocracy.ca/en
7. The OGP currently has subnational pilot projects – http://www.opengovpartnership.org/how-it-works/subnational-government-pilot-program

8. Provinces and Territories in Canada, for example, have very rich information assets as this is the jurisdiction that delivers health, transportation, social services and education, and who will administer and settle refugees even though this was a national decision. Arguably the administrative data and information from this level of government are those which are close to and relevant to citizens.

9. SEIA is a very formal mechanism also tied to overseas development aid such as the World Bank – http://www.ifc.org/performancestandards

10. The checklist includes commitments to: access to information, aid transparency, anti-corruption, budget transparency, citizen participation, e-government, legislative openness, natural resources, open data, public procurement, public service delivery and subnational governance.

11. http://legacy.earlham.edu/~peters/fos/timeline.htm

12. http://oad.simmons.edu/oadwiki/Timeline

13. http://media.transparency.org/maps/cpi2015-470.html

14. Snippe: http://www.transparency.org/assets/data/cpi2015/cpi-data.json

15. http://www.access-info.org/

16. http://www.opengovstandards.org/

17. https://publicadministration.un.org/egovkb/Portals/egovkb/Documents/un/2014-Survey/Chapter3.pdf

18. http://www.iap2.org

19. Modified from https://c.ymcdn.com/sites/www.iap2.org/resource/resmgr/Foundations_Course/IAP2_P2_Spectrum.pdf

ACKNOWLEDGEMENTS

Tracey's research for this chapter and book was funded by a European Research Council Advanced Investigator grant, The Programmable City (ERC-2012-AdG-323636).

REFERENCES

Anderson, B. (1991) *Imagined Communities: Reflections on the Origin and Spread of Nationalism.* London: Verso.

Brownlee, J. and Walby, K. (eds) (2015) *Access to Information and Social Justice Critical Research Strategies for Journalists, Scholars, and Activists.* Winnipeg: ARP.

Cartwright, W., Peterson, M. and Gartner, G. (2007) *Multimedia Cartography*, 2nd edition. New York: Springer.

Cavoukian, A. (2010) *Access by Design*, Information and Privacy Commissioner of Ontario, Toronto. Available at: https://www.ipc.on.ca/images/Resources/accessbyde sign_7fundamentalprinciples.pdf (accessed 27 July 2016).

Cavoukian, A. (2011) *Privacy by Design*, Information and Privacy Commissioner of Ontario, Toronto. Available at: https://www.ipc.on.ca/images/Resources/7foundati onalprinciples.pdf (accessed 27 July 2016).

Craglia, M., Leontidou, L., Nuvolati, G. and Schweikart, J. (2004) 'Towards the development of quality of life indicators in the digital city', *Environment and Planning B: Planning and Design*, 31 (1): 51–64.

Crampton, J. (2001) 'Maps as social constructions: power, communication and visualiza-
    tion', *Progress in Human Geography*, 25 (2): 235–52.
Elwood, S. and Leszczynski, A. (2012) 'New spatial media, new knowledge politics',
    Transactions of the Institute of British Geographers, 38 (4): 544–59.
Feenberg, A. (2011) 'Agency and citizenship in a technological society', lecture pre-
    sented to the course on digital citizenship, IT University of Copenhagen. Available at:
    https://www.sfu.ca/~andrewf/copen5-1.pdf (accessed 27 July 2016).
Francoli, M. (2014) *Independent Reporting Mechanism: Canada Progress Report: 2012–2013*,
    Open Government Partnership. Available at: http://www.opengovpartnership.org/sites/
    default/files/Canada_final_2012_Eng.pdf (accessed 27 July 2016).
Francoli, M., Ostling, A. and Steibel, F. (2015) *From Informing to Empowering: Improving
    Government-civil Society Interactions Within OGP*, Open Government Partnership, July.
    Available at: http://www.opengovpartnership.org/country/case-study/informing-
    empowering-improving-government-civil-society-interactions-within-ogp-0
    (accessed 27 July 2016).
Fremlin, G. and Robinson, A. (1998) 'Maps as mediated seeing', *Cartographica*, 35 (1/2):
    Monograph 51.
Garsten C. and de Montoya, L. (eds) (2008) *Transparency in a New Global Order: Unveiling
    Organizational Visions*. Cheltenham: Edward Elgar.
Government of Ontario (2014) 'Open by default – a new way forward for Ontario'.
    Available at: https://www.ontario.ca/document/open-default-new-way-forward-
    ontario (accessed 27 July 2016).
Harley, J.B. (1989) 'Deconstructing the map', *Cartographica*, 26 (2): 1–20.
Holzner, B. and Holzner, L. (2006) *Transparency in Global Change: The Vanguard of the
    Open Society*. Pittsburgh, PA: University of Pittsburgh Press.
Kitchin, R., Lauriault, T. and McArdle, G. (2015) 'Knowing and governing cities through
    urban indicators, city benchmarking and real-time dashboards', *Regional Studies, Regional
    Science*, 2 (1): 6–28.
Kitchin, R., Lauriault, T. and McArdle, G. (2016) 'Smart cities and the politics of urban
    data', in S. Marvin, A. Luque-Ayala and C. McFarlane (eds), *Smart Urbanism: Utopian
    Vision or False Dawn?* London: Routledge. pp. 16–33.
Lauriault, T. and O'Hara, K. (2015) '2015 Canadian election platforms: long-form census,
    open data, open government, transparency and evidence-based policy and science',
    SSRN. Available at: http://ssrn.com/abstract=2682638 (accessed 27 July 2016).
Lord, K.M. (2006) *The Perils and Promise of Global Transparency: Why the Information
    Revolution May Not Lead to Security, Democracy, or Peace*. Albany, NY: State University of
    New York Press.
McGonagle, T. and Donders, Y. (eds) (2015) *The United Nations and Freedom of Expression
    and Information: Critical Perspectives*. Cambridge: Cambridge University Press.
O'Hara, K. and Dufour, P. (2014) 'How accurate is the Harper Government's misinfor-
    mation? Scientific evidence and scientists in federal policy making', in G.B. Doern
    and C. Stoney (eds), *How Ottawa Spends, 2014–2015: The Harper Government – Good
    to Go?* Montreal: McGill-Queen's University Press. pp. 178–92.
Parks, W. (1957) 'The open government principle: applying the right to know under the
    constitution', *The George Washington Law Review*, 26 (1): 1–22.
Peterson, M.P. (1995) *Interactive and Animated Cartography*. Englewood Cliffs, NJ: Prentice
    Hall.
Porter, T. (1995) *Trust in Numbers: The Pursuit of Objectivity in Science and Public Life*.
    Princeton, NJ: Princeton University Press.

Pulsifer, P.L. (2005) 'The cartographer as mediator: cartographic representation from shared geographic information', in Taylor, F. (ed.), *Cybercartography: Theory and Practice*. Amsterdam: Elsevier. pp. 149–79.

Sheedy, A. (2008) 'Handbook on citizen engagement: beyond consultation', Ottawa: Canadian Policy Research Network. Available at: http://www.cprn.org/documents/49583_EN.pdf (accessed 27 July 2016).

Star, S. and Griesemer, J. (1989) 'Institutional ecology, "translations" and boundary objects: amateurs and professionals in Berkeley's Museum of Vertebrate Zoology, 1907–39', *Social Studies of Science*, 19 (3): 387–420.

Straumann, L. (2014) *Money Logging: On the Trail of the Asian Timber Mafia*. Zurich: Bergli Books.

Sui, D.Z. and Goodchild, M.F. (2001) 'Are GIS becoming new media?', *International Journal of Geographical Information Science*, 15 (5): 387–90.

Wilson, M. and Stephens, M. (2015) 'GIS as media?', in S. Mains, J. Cupples and C. Lukinbeal (eds), *Mediated Geographies and Geographies of Media*. Berlin: Springer. pp. 209–21.

Worthy, B. and Hazel, R. (2014) 'The impact of the Freedom of Information Act in the UK', in N. Bowles, J.T. Hamilton and D.A. Levy (eds), *Transparency in Politics and the Media*. New York: I.B. Taurus and Co. pp. 31–45.

19
PRODUCING SMART CITIES

MICHAEL BATTY

CHANGING CONCEPTIONS OF THE CITY

Just as our cities have become less polluted, greener and, at least in the West, more prosperous, our interest in them has dramatically accelerated. Big cities, in particular, are the flavour of the month. They attract the young and the talented, the cool and the creative. They are fast becoming the places where the future is being invented through new information technologies (IT), new lifestyles, and new modes of entertainment, consumption and work. Into this milieu has come the idea that cities can be automated in the same way that industrial systems were automated in earlier generations, but this time around it is the collective functions of cities that are being recreated through new IT and new ways of communication. Many of these technologies constitute new spatial media, given their production, processing and dependence on vast quantities of spatial big data and heavy utilisation of spatial interfaces such as maps and spatialisations. In short, we now speak of the 'smart city', wherein cities are endowed with some intelligence in their functions, which in the past have been somewhat dumb in their operations. In one sense, our new-found interest in cities cannot be separated from developments in digital technologies. Advances in computation and communications are leading to many new forms of social media that in turn are augmenting and changing the functions that continue to dominate the way our cities work. In this sense, the development of the smart city is just another wave in the development of information technologies that have spread out from science to business and government over the past 50 years, and in the past 20 years from the individual to the collectivity, culminating in what some are calling a new 'operating system' for the city.

There is a good deal of hype involved in this notion that our cities are suddenly becoming smart simply because our IT have continued to spread out and diffuse into all corners of our lives. But once we cut through the rhetoric, the smart city essentially embraces the use of digital technologies that are being used to automate routine day-to-day functions that enable populations to communicate with each other and with the artefacts of the built environment that are used to make the city work. Moving people, goods, information and ideas using technologies that complement traditional modes of communication is key, and providing routine government and non-market services using IT such as those related to location, and engaging citizens in various public domains, are central to the smart city. To do this, the entire panoply of IT devices underpins these patterns of delivery and communication. In the past decade, personal devices that we use to communicate with one another and with artefacts have become central to such operations, the exhaust from which is providing enormous quantities of data – big data (vast quantities of which are georeferenced) – that we are beginning to use to make sense of the way these augmented functions are changing the city.

Ways of using these new devices to capture and spread new media are enriching and changing the way we communicate socially, and such social media are providing diverse ways to supplement many traditional functions involving work and entertainment. We are beginning to provide new interfaces to the way cities operate by synthesising these data visually into dashboards that provide an instant picture of city life (see Chapter 7), and for these we are evolving many new forms of urban analytics (see Chapter 14) enabling us to produce more insightful analyses and predictions of how cities can be improved in terms of their functioning. At least that is the intention and the hope. So to summarise, new forms of digital communications, new services delivered digitally, and new communities fashioned from social media constitute the smart city, made possible by all-pervasive digital devices that are either embedded into the fabric of our cities (producing real-time streams of sensed data) or operated in mobile fashion by ourselves. These are all generating data that a new array of analytics is beginning to make sense of, directed towards a new type of science that is enabling us to better understand cities in general and the smart city in particular.

PROGRESS TOWARDS A NEW SCIENCE

Until quite recently our thinking about cities in formal ways has been based on their development to an assumed equilibrium over quite large time spans. In fact, we tend to articulate cities as we see them at a single point in time. Consequently, many of our theories, about how their functions generate the physical forms we see and how their populations behave with respect to the way land is used, describe a timeless world in which locations and forms are explained as if they were suddenly created at that instant. Most of our formal models suggest a world in equilibrium in contrast to the very large body of urban studies focusing on explaining the city more descriptively in verbal terms, which does examine its dynamics but, again, usually over the longer term. Only very recently has the notion that we can explain the routine short-term dynamics of the city more formally come onto the agenda. This is as much through the emergence of big data as it is through the shortening of our attention spans in a world that is ever faster and seemingly more responsive.

At present, our urban theories provide a mixture of those that deal with the short term and the long term, with cities in equilibrium as well as out of equilibrium

(Batty, 2016). There is, however, no sense in which there is an integrated scientific perspective making sense of all this variety. There are, though, some features of what we know about cities that are beginning to forge such a new science. Central to this is the notion that cities are all about flows – about interactions and relationships – which underpin the patterns that we observe with respect to locations, places and the way we carve the city up into different spaces. To understand location, interactions, movements, communications and so on is now deemed to be essential (Batty, 2013). This is a long-standing idea but it is surprising that we have not taken it more to heart. With the rise of social media and with movement and communications being recorded digitally through everything from email to smartcards, the idea of the city as a communications medium has become critical in ways that commentators such as Richard Meier (1962) and Melvin Webber (1964) anticipated so acutely and so long ago.

Data that are streamed or captured by mobile or fixed sensors are largely unstructured, but in cities they tend to show great heterogeneity and diversity. They are usually highly disaggregate if collected and associated with individuals using smartcards or phones. In this sense, they provide rather good material for micro- and agent-based simulations. Because they are temporally dynamic in their basic form and usually extensive, they require different kinds of analytics and modelling from that associated with much more structured, smaller datasets that are collected at a cross-section in time from population censuses and the like. Multivariate statistics, which had developed rapidly from over half a century ago, are increasingly inappropriate on which to base such analytics, and new forms of data mining focusing on extracting patterns from machine learning are being evolved. Models have become much more disaggregate and new computer methods based on massively parallel distributed processing are now required to make sense of such data. Visualisation is increasingly important and this focus has changed from long-term and cross-sectional simulation to much more routine operations and functions in cities. For example, disruption, crime and health emergencies, occurring daily, are of more concern with respect to modelling and prediction. All of this is being painfully grafted onto the overall structure of knowledge in this domain, while some of the old certainties with respect to how cities are structured are rapidly disappearing. These certainties have gone as a new paradigm focused on complexity has propelled forward formal frameworks associated with agent-based modelling, micro-simulation, serious games, network processes and cellular automata.

It is going to take some time for this new science to be constructed, and like all knowledge it is contingent on the times in which it is being invented and discovered. Cities are getting ever more complex, and IT which are being fast embedded into their fabric are changing the very way in which we behave. There is now a clear imperative that suggests that as we observe and probe the city, we change it, just as we do in the more esoteric branches of physics dealing with quanta. Indeed, the models and analytics that we are using are changing the very systems that we are seeking to understand and manipulate. In the smart city, everything is contingent and changeable. To give some sense of this, I set out three key dimensions of the future city: integrated platforms or operating systems for cities; communications and movement; crowdsourcing and citizen science (see Chapter 12). All of these are being informed by new IT that are central to how we behave and interact with one another in the smart city.

ADAPTING NEW TECHNOLOGIES

INTEGRATED PLATFORMS AND CITY OPERATING SYSTEMS

The most optimistic perspective on the smart city suggests that we can apply quite literally the idea of the operating system to organising many functions that a city generates. This, to an extent, is fanciful for it assumes we know and understand the range of functions and how they relate to each other, and can integrate them somewhat seamlessly into a form where they are interoperable. Our experience hitherto with large IT systems is not good, particularly where strong top-down imperatives are used to build such systems, as for example in national agencies – in Britain the National Health Service is typical and indeed apocryphal. Many of the largest IT companies, such as IBM, which concentrate on computer services and software, and Cisco, which dominates networks and switching, have implicitly endorsed this idea, and in the flurry of planned technological new towns such as Masdar and Songdu, there are demonstrations of how such systems might be developed. In fact, Living PlanIT are developing the 'PlanIT Urban Operating System' originally for an area east of Porto in Portugal, but this is, I feel, more a metaphor for integrating IT than a realisable proposal.

The difficulties of coupling systems do not pertain just to interoperability but also to questions of data integration and even the styles of models and simulations that pertain to different sectors. For example, unless there are common keys between datasets – of which there are very few in cities, geocodes being the most obvious – it is impossible to integrate anything. Transport for London can measure demand in real time from its Oyster smartcard, which records all tap-ins and tap-outs of its travellers, but it is impossible to link this to the supply of trains, which is also known at precise locations and time instants from their operating systems. Within a Tube station travellers cannot be tracked – it is illegal – and in any case the Wi-Fi is not sufficiently detailed and may never be good enough to track 'what people' get onto 'what trains'. Currently it is impossible even with state-of-the-art equipment and data to cement this linkage. If Transport for London is unable to do this, no one else can, and this kind of circumstance is writ large across our cities.

A more modest aim for cities is integration that is built from the bottom up – in what are increasingly called platforms (Grech, 2015). These are digital forums for interfacing a variety of application programming interfaces (APIs) – pieces of software that can be queried, linked and coupled to each other, but ordered in certain ways that reflect the purpose of the integration. Generic software is being invented in this way so it can take account of very different pieces of established and not-so-established software and data that can be linked into platforms that offer the generic software as a service. One of the big differences between seeing cities as platforms or as operating systems is not just simply one of ambition, but one of modularity where the components of the platform can be incessantly replaced with better and/or different versions of the same, producing the possibility of continual disruption as new software is created from the old that is being destroyed. This model is one that appears to be endemic to IT as witnessed in the great transition from hardware to software, from large to tiny computers, and from remote individual access to all pervasive networking. IT are increasingly subject to Schumpeter's notion of 'creative destruction'; no longer within the cyclical Kondratieff wave of 75 years' duration but now being almost continuous in its transformation (Batty, 2015).

To date, no operating systems or platforms have actually been built that embody the notions of the smart city, unless one counts ad hoc attempts at stitching available software and sensor systems together in a loosely coupled manner. The rhetoric is more one of metaphor than of actuality, but what have been developed are portals where data are drawn together, either in dashboard-like interfaces, one of which the Centre for Advanced Spatial Analysis (CASA) has constructed for London (see Figure 19.1) (O'Brien et al., 2014), or in physical locations built around the idea of city control rooms. Most big cities have such rooms for specialist purposes where manual and automatic controls of systems such as water, electricity, traffic and so on are operated. Proposals for city-wide services such as IBM's Rio de Janeiro implementation show what is possible (Kitchin et al., 2015). The basic problem remains, however. First, the data and control that are possible are not easy to integrate across different systems, notwithstanding the fact that such portals are still of interest in providing a big picture. And second, many of the key indicators that we need to visualise with respect to how cities are performing cannot be sensed automatically, for they depend on social and economic data such as housing, ownership and tenancy, market

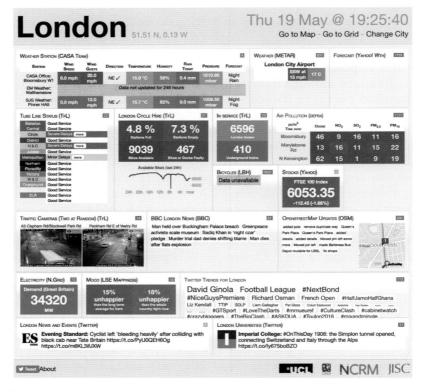

FIGURE 19.1   The London Dashboard – real-time data pertaining to the city

prices, unemployment and so on. These data are slowly coming but they may never be on the same frequency as more physical and travel-based data, for they are only likely to be released on monthly, perhaps (at best) weekly, cycles. This changes the picture quite radically of how a city is performing from the minute-by-minute displays that characterise the current generation of dashboards.

COMMUNICATIONS AND MOVEMENT IN THE CITY

If you examine the dashboard in Figure 19.1, you will see many inputs of data that pertain to how people, materials and ideas are moving within the city – frequencies of subway (Tube) services, what is trending on Twitter and so on. Cities are places where people come together to engage socially and economically, to share and divide their labour, to add value to what they do to sustain themselves. Cities are quintessentially marketplaces for ideas and material transactions. If we could all interact at the same physical location, we would not require any physical movements to enable these transactions to take place, but given the exigencies of physical structure and the fact that as human beings our own personal space is geometrically distinct, many different kinds of movements take place in cities to enable cities to function effectively. Much urban planning attempts to reduce the scale and volume of these movements although this is quite a controversial view in that some argue that movement in fact is necessary for the good functioning of cities and that, in the modern world, globalisation essentially projects this idea to its extreme. The argument is one that relates to density, compactness, sprawl and city size, over which there is little agreement.

We have good data on how public transport is being automated in London and this is increasingly the case in many big cities, where smartcards are replacing more manually based charging systems. Oyster Cards are used in London to record where and when and who taps in and out of the Tube and other train network stations. Buses are trickier as single drivers mean that only tap-in can be monitored. In fact in Singapore the EZ-Link system records tap-in and tap-out for trains and buses, but this difference reflects cultural mores, which are important in understanding the extent to which cities can be automated. This kind of dataset provides just one perspective on physical travel. Because it is unstructured and relates to places where people enter and exit a transit system, it is very much an incomplete picture of how people move from fixed locations that relate to places where they undertake activities such as education, entertainment, work and so on. It needs to be integrated with other datasets so it can be given structure.

This is a very basic point with respect to many new sources of travel data. For example, mobile phones log where a call is made from and to, and these can be extracted from the call detail record (CDR). With judicious analysis of times when such calls are made and the relative longevity of time spent in fixed places, prudent estimates of the correlation between physical travel patterns and the flow of calls can be made. In fact, so far, although these data look promising, they are very hard to integrate and improve without independent data that can be related to them. Doing the same with Twitter and other social media data is equally problematic. The key challenge is shown in Figure 19.2, where we have data on origins, destinations and trip patterns from small areas in Greater London for travellers who use a variety of Tube, road, bus, rail, bike and walk modes to get to work (Figure 19.2a, c, b), and

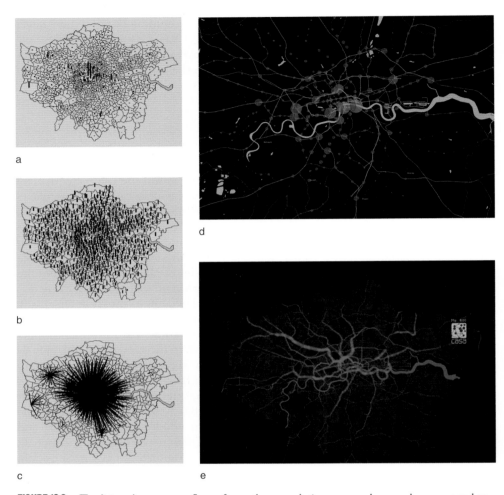

FIGURE 19.2 Traditional transport flows from the population census by wards compared to tap-ins and tap-outs from Transport for London's Oyster Card data. (a) Origin employment (2001) by wards, (b) destination working population (2001) by wards, (c) trips (2001) greater than 200 persons per ward to ward flow, (d) tap-ins and tap-outs at 8:00am on the London Tube (November weekday 2010), (e) flows of tap-ins/-outs assigned to the tube network

also the flow of data on the Tube at the peak hour where we assume that travellers take the shortest routes between the points where they tap in and tap out using their smartcards (from the Oyster Card data in Figure 19.2e). These two sets of patterns might be supplemented by other detailed flows by mode and volumes of traffic on the highway from loop counters. The picture we are able to get is highly variegated with respect to error and ambiguity and as yet we do not even know if these sorts of data coming from real-time streaming in the smart city are going to be useful for the traditional practices of transport planning.

These transport flows are material movements in that they are measured here in persons but they could be commodities, freight and so on. The smart city is also being fast automated with respect to flows of information measured in terms of communications using social media, financial transactions and email/web-related traffic. There are many diverse sources of such data; to illustrate one, the Spanish BBVA Bank have produced data from their credit card transactions which have been animated to show spending patterns over short periods of time. The movie produced by the Senseable City Lab at MIT shows such flows over Easter 2011,[1] and Figure 19.3 shows two infographics from data that pertain to financial flows that are being used to figure out where people shop and how much money they spend. Figures 19.3a, b and c show origin and destination maps of sales and a

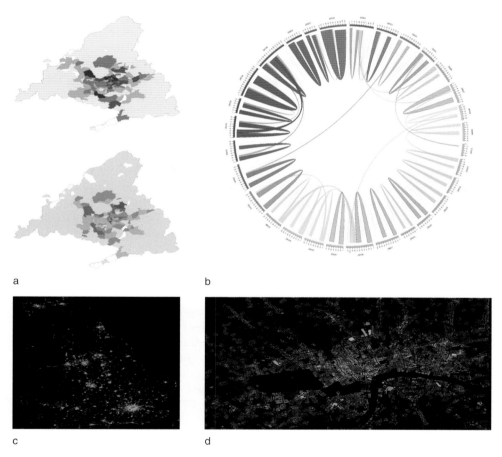

a

b

c

d

FIGURE 19.3 Credit card data for retail sales in Madrid and Foursquare check-ins from the social media correlated with retail sales in the UK and London. (a) Origins and destinations of sales in Madrid with (b) a circular visualisation using Circos software of flows that show most inside the relevant zonal area, (c) density of Foursquare check-ins in the UK, (d) similar at the scale of retail locations in Central London

circular flow graph of the flows of credit card transactions aggregated to 32 locations in Madrid. Figures 19.3c and d show the clustering of Foursquare check-ins
defining shopping patterns in the UK and Central London, which can be used as
a proxy for where people shop.

CROWDSOURCING, OPEN DATA AND VALUE THROUGH ANALYTICS

So far it is clear that the exhaust from real-time control, which is one of the missions
of the smart city movement, provides many new data sources that inform us faster
and more precisely than ever before, but this also raises important problems of
representativeness. In fact, it is fairly obvious but needs reinforcing that it is the
World Wide Web – the internet or net of networks – which is the new glue that
is tying us and our objects and artefacts all together. It is an internet of things
that is driving the new ways in which we can share data, simulations and forecasts
among ourselves with the prospect that all of us can engage in altering those very
data, models and forecasts which in the past remained the prerogative of the very
few. New ways of opening data are now possible, but only because of the web.
So much government data are now in the public domain, such data having been
deregulated, that this is enabling many new software products to be developed that
add value to such data (see Chapter 9).

Open data, however, is simply the tip of the iceberg with respect to new and big
data that are being sensed, collected and collated. Data that are collected by ourselves
through various forms of crowdsourcing – mobilising the power of the crowd who
invariably are responding to mandates to provide data and opinions in some semi-
structured fashion – are becoming ever more significant. Much of the data, sensed
in this way through online questionnaires, are providing data beyond that which are
sensed in real time. These data are controversial as they reflect interpretations and
opinions that cannot often be laid bare, but some are quite objective. OpenStreetMap
(OSM) – built worldwide by a generation of those interested in location and place
and tagging using global positioning systems (GPS) on smartphones – is an excellent case in point. Much of the data in volume terms are as good as those collected
by national mapping agencies. There are many such examples of crowdsourcing, but
for it to be really effective, it needs a broadcast medium to tell the crowd that there
are those who want to elicit their opinions. Some of this is done by the internet
itself and by the use of social media, but the best is where there is a medium like TV
that is widely accessed and through which the crowd can be mobilised. An example
that our group has worked upon was based on local TV media running a campaign
to elicit responses to key problems of social behaviour and to enable TV watchers
to log and record their responses to the relative importance of social problems by
recording their responses and their locations. Work with the BBC local news programme *Look East* elicited many thousands of responses across the eastern region
of England concerning the preponderance of key problems such as drunkenness,
stolen cars, noise, pollution and so on (Hudson-Smith et al., 2009). In Figure 19.4,
we show an example of this with a shot from the broadcast TV medium alongside

FIGURE 19.4   Crowdsourcing: a broadcast medium to reach TV viewers

the map that was produced over a short period of time, it being continually updated as more and more watchers responded.

There are many different variants on this notion of collecting data from the crowd, which range from routine lower-level responses, to relatively unambiguous queries, to scientific sourcing where users are enticed to solve problems, discover new ideas and present their insights into key controversies in a structured way. One of the things that has not been introduced so far is any review of the range of theories and analytical techniques that are being fashioned to deal with these kinds of data. New statistical models are clearly needed as noted with respect to dealing with datasets whose structure and volume are qualitatively different from previous data volumes that were much smaller. New methods of data mining are important that are able to extract deep structures and structures that vary widely across any single dataset. In short, the bigger the dataset, the more variegated the structures that are possible within it, and our models and techniques need to be adapted to take all this into account. Respondents are asked answer questions about social behaviour in their own locality and to record such responses using CASA's online mapping system MapTube.[2]

You might anticipate that a new science of cities is developing by building on what has gone before, where an older science produced analytics well adapted to our prior understanding of cities and the needs in their planning. Nothing could be further from the truth. Cities, like all human systems, are continually in flux and their complexity appears to be outstripping our abilities to make sense of them. The previous section on transport flows provided a glimpse at the past generation of models that deal with flows, but the new data which are streamed in real time change the focus quite radically, and as yet we have not been able to integrate models that deal with different temporal resolutions and intervals. New models and new forms of analytics that go beyond data mining are urgently needed and although there is some prospect that more disaggregate models – agent-based models – might hold the key to integrating different spatial and temporal scales, the field is wide open for new insights and new theories that make sense not only of the smart city, but also of the wider context of spatial media of which the smart city is a part.

CHALLENGES FOR STAKEHOLDERS AND CITIZENS

A final feature of the smart city relates to those who live in it – ourselves, acting as citizens and stakeholders. The web is bringing a wide variety of services to us that we have traditionally travelled to consume or at least collect, as well as services that are entirely new and a product simply of the fact that we can now access them using IT. Every piece of information that traditionally required us to manually complete paperwork can now be delivered by the internet and processed online. In principle, all bureaucratic transactions can currently be completely recorded, but the data volumes are huge, and as in earlier times when these tasks were not automated, organising and understanding their wider import is as significant as it ever was. Elsewhere in this book, there is discussion of the extent to which this new media is creating a citizen science (see Chapter 12), new forms of participation (see Chapter 18), questions of privacy and confidentiality in data (see Chapter 22), as well as the limitations and potential of all these new ways of working. These are all part and parcel of the smart city.

The challenges we face in the development of smart cities are enormous. Large IT companies who see the smart city as just the next phase of the rollout of their business have found a new mission in wrapping up traditional ways of making money, seemingly invoking a mission with social objectives. Much of the smart-city rhetoric is phrased in terms of achieving a better quality of life for citizens, but the debate is strangely silent about questions of segregation, inequality and poverty, and tends to focus more on accessibility and economic opportunity. This is where the wider constituency of academics, policy-makers and citizens at large have a critical role to play. In leading the debate about how new information technologies can be adapted, ways in which both public as well as private goals that the community and its individuals might achieve in the wider context of such automation can be articulated and reached.

ACKNOWLEDGEMENTS

Ollie O'Brien devised the original London dashboard and Steven Gray extended the media both within it and the way it displays. Jon Reades was instrumental in developing our work with Transport for London's Oyster Card data; thanks to Joan Serras for the work with the BBVA credit card data for retailing in our EUNOIA project, and Vassilis Zachariadis for the retail model that is using Twitter and Foursquare data to build proxies for retail attractions in our INSIGHT project. Richard Milton devised MapTube and was central to the *Look East* crowdsourcing project.

NOTES

1. https://www.youtube.com/watch?t=13&v=8J3T3UjHbrE
2. http://www.maptube.org/

REFERENCES

Batty, M. (2013) *The New Science of Cities*. Cambridge, MA: MIT Press.

Batty, M. (2015) 'The sixth Kondratieff is the age of the smart city', Spatial Complexity. Available at: http://www.spatialcomplexity.info/files/2015/05/Draft-Chapter-2015.pdf (accessed 27 July 2016).

Batty, M. (2016) 'Cities in disequilibrium', in J. Johnson, A. Nowak, P. Ormerod, B. Rosewell and Y.-C. Zhang (eds), *Non-Equilibrium Social Science*. Berlin: Springer.

Grech, G. (2015) 'Cities as platforms', *Tech Crunch*, 7 August. Available at: http://techcrunch.com/2015/08/07/cities-as-platforms/ (accessed 27 July 2016).

Hudson-Smith, A., Batty, M., Crooks, A. and Milton, R. (2009) 'Mapping for the masses: accessing Web 2.0 through crowdsourcing', *Social Science Computer Review*, 27 (4): 524–38.

Kitchin, R., Lauriault, T.P. and McArdle, G. (2015) 'Knowing and governing cities through urban indicators, city benchmarking and real-time dashboards', *Regional Studies, Regional Science*, 2: 6–28.

Meier, R.L. (1962) *A Communications Theory of Urban Growth*. Cambridge, MA: MIT Press.

O'Brien, O., Batty, M., Gray, S., Cheshire, J. and Hudson-Smith, A. (2014) 'On city dashboards and data stores', a paper presented to the Workshop on Big Data and Urban Informatics, 11–12 August 2014. Chicago, IL: University of Illinois. Available at: http://urbanbigdata.uic.edu/proceedings/ (accessed 27 July 2016).

Webber, M.M. (1964) 'The urban place and the nonplace urban realm', in M.M. Webber, J.W. Dyckman, D.L. Foley, A.Z. Guttenberg, W.L.C. Wheaton and C. Bauer Wurster (eds), *Explorations into Urban Structure*. Philadelphia, PA: University of Pennsylvania Press. pp. 79–153.

20

SURVEILLANCE AND CONTROL

FRANCISCO KLAUSER AND SARAH WIDMER

The aim of this chapter is to highlight and problematise the surveillance dynamics inherent in the contemporary proliferation of new spatial media. In addressing this problematic, the chapter is structured in two main parts. The first section provides a broad outline of the surveillance potential – and functioning through surveillance – of spatial media. We also discuss the key issues arising from the increasing digitisation and interconnection of various forms of new spatial media/spatialised data collection and analysis, and we critically assess the roles and positions of two key stakeholders connected with these issues: the individual user of spatial media, and the technical experts involved in coding everyday life into software. The second section of the chapter then focuses on three main aspects that characterise the surveillance dynamics implied by new spatial media: personalisation, interconnection and anticipation. The chapter's conclusion argues for further analysis and conceptualisation of the increasing flexibility of contemporary governing through code with respect to new spatial media.

SURVEILLANT POTENTIAL OF NEW SPATIAL MEDIA

By creating spatialised informational environments that mediate daily-life practices, new spatial media rely fundamentally on systematic forms of data gathering and analysis (Green and Smith, 2004; Farman, 2011; de Souza e Silva and Frith, 2012). Put differently, new spatial media are inherently surveillant in functioning and implication, if we understand surveillance as the 'practices and techniques aiming at the focused, systematic and routine attention to personal details for purposes of influence, management, protection or direction' (Lyon, 2007: 14). For example, in order for Google Maps to provide its users with a map corresponding to their location, it has to gather positioning information from their devices. This is complemented with other types of information that Google collects and analyses (users' search history, content of their Gmail account, etc.) in order to personalise its cartographic service.

Thus, the main difference between such applications and traditional spatial media (such as a map or a *Lonely Planet* travel guide) lies in the fact that Google can collect and process information on who and where we are, what places we are interested in, combine that information with data from its other services, and do so each time the application is used. Therefore, by using these applications, we produce what Kitchin and Dodge (2011: 90–91) have named a 'capta shadow', a digital shadow of ourselves, which reflects our locations, preferences, practices and relations – a shadow that is growing and persistent over time.

If surveillance is the very condition and price to pay for spatial media to achieve their proclaimed benefits – simplify everyday life, anticipate individual needs, optimise specific activities, etc. – the increased possibilities of knowing and tracking daily-life activities raise a series of critical issues. These range from the effects on privacy and social trust, and the lack of accountability and transparency, to the risks associated with information sharing, the potential of social discrimination, and the role of private interests in the design and use of spatial media applications (Cost Action IS0807, 2008). To further pursue this reflection, it is worth pointing towards three sets of issues or problems in particular.

PRIVACY AND DATA-PROTECTION CONCERNS

The first set of problems revolves around privacy and data-protection concerns, related to issues arising from the commercial and political exploitation of data accumulated by new spatial media (see Chapter 21). As shown by the recent National Security Agency (NSA) leaks for example, government agencies can have a strong interest in collecting data about their citizens or foreign organisations/citizens for purposes ranging from tax-fraud suspicion to the detection of terrorist activities (Albrechtslund, 2012). Furthermore, data generated by new spatial media can also have a considerable commercial value, allowing companies to profile consumers and to market products accordingly (see Chapter 17). Thus although these data are often used in aggregated form, their production and storage on external servers significantly increases the scope of what is recorded and potentially searchable about one person, thus undermining individual privacy (Lessig, 2006: 202).

Yet, despite the dangers and ethical dilemmas implied, the 'omni-memory' of spatial media can also be experienced in positive ways by their users. As shown in more detail elsewhere (Frith, 2015; Widmer, 2015a, 2015b), many users of locative media perceive the storage of their locational data as something practical, allowing them to delegate their memories of the places they have visited to technology. This outsourcing of one's memory is part of a broader trend of 'self-tracking', consisting for instance in the monitoring of one's sleep, sports performance, number of steps taken, etc. through the use of wearable devices (such as fitness trackers) and various smartphone applications (Klauser and Albrechtslund, 2014). These self-tracking practices challenge the common representation of top-down surveillance, where the individual would merely be the passive object of a monitoring conducted by governments or private corporations. Here, on the contrary, individuals are active initiators of their own surveillance – what Mann et al. (2003) term 'sousveillance' – in deciding themselves which application and spatial media to use. Yet this freedom to decide is informed and governed on all kinds of levels and in all kinds of ways, including financial incentives, information campaigns, advice generated by software, etc. Together, these mechanisms form a mode of regulation that does not work in a disciplinary way (through rigid prohibitions or prescriptions), but which acts on the user's own desire to benefit from the data-derived advantages offered by new spatial media.

SOCIAL SORTING

The second set of problems can be subsumed under the heading of 'social sorting' – that is, the categorisation and differential treatment of individuals based on their calculated worth and eligibility (Graham, 2005), for example algorithmically processing data to determine whether a person should be given a job or loan or tenancy. A key aspect of spatial media is not merely data gathering and transfer, but information processing and analysis through software to generate automated responses (Thrift and French, 2002; Kitchin and Dodge, 2011). Put differently, at their very core, spatial media rely on the coding of social life into computer algorithms that automatically perform tasks (Graham, 1998; Haggerty and Ericson, 2000; Lyon, 2007).

As shown by a range of scholars, the implied processes of data analytics are never neutral, whether the collection, classification and processing of data aim at greater efficiency, convenience or security (Thrift and French, 2002; Graham, 2005; Lyon, 2007). Instead, they depend on technologically mediated codes that are used to assess and orchestrate everyday life. These codes constitute often-invisible processes of classification and prioritisation, which may affect the life-chances of individuals or social groups in ways that are frequently opaque to the public and which easily evade conventional democratic scrutiny. As Graham et al. (2013: 470) note with regard to spatial media, 'the apparently straightforward relationship between content sought and content displayed is usually mediated by complex algorithms that tailor information based on the interactions of several factors.' In other words, on the basis of the data that are collected and analysed, algorithms shape the visibility and the invisibility of content on spatial media, channelling users' choices and decisions about where to go. When, for example, algorithms personalise information in order to match users' interests or preferences, the resulting tailored maps and spatial recommendations produce filtered informational landscapes where the user only sees what resembles them the most (Pariser, 2011). Those filtered informational landscapes reinforce socially and demographically homogenous 'communities of like-minded people' (Graham, 2005: 571) and contribute to the splintered geographies that characterise our contemporary societies. New spatial media thus raise a series of critical power issues arising from the codes' use to assess people's profiles, risks, eligibility, and levels of access to various spaces and services, thus instilling a new kind of 'automatically reproduced background' in everyday life (Thrift and French, 2002: 309).

INTERESTS BEHIND CODE

Following from this, the third set of problems concerns the question of who defines and controls the computer algorithms that allow new spatial media to work. In recent years, an increasingly detailed body of work has shown that novel software-mediated techniques of regulation and control further exacerbate the reliance on the role of private actors and technical expertise in defining, optimising and managing the 'control by code' (Lyon, 2007: 100) of urban systems and services. Managing, ordering and governing, in this context, means to make use of the mediating techniques and mechanisms involved in coding everyday life into software. Thus, authority derives from the expertise necessary for the design and use of computer algorithms needed to control, sort and associate the masses of data generated and processed. Giving certain (private) parties more weight challenges traditional modes of governance in which the management and control of individuals were the exclusive responsibility of the nation-state (Cost Action IS0807, 2008: 19). It also raises the critical question of how commercial goals, particularly when they intersect with public interests, situate themselves in relation to wider considerations such as democracy, accountability and efficiency. Yet, despite the significant ethical issues raised by corporation involvement in the regulation and control of everyday life, there is a dearth of theoretical and empirical work on questions of how, by whom and for what reasons such systems are being developed and deployed.

POWER DYNAMICS IMPLIED BY NEW SPATIAL MEDIA

One way to move beyond this broad discussion of the surveillance implications of new spatial media is to focus more specifically on the regulatory dynamics implied by the technologies' functioning through data accumulation and data analytics. If we are to understand the surveillance enabled by spatial media, we need to foreground the basic rationalities of power inherent in the specific forms of regulation and control that arise from their use. Here, we focus on three such rationalities: personalisation, interconnection and anticipation.

PERSONALISATION

Whether we are using Google Maps for directions, Foursquare for recommendations, or getting navigational indications from Google Now, geolocational services differentiate their content depending on what their algorithms understand about where we are (location- and context-aware apps), who we are (user-aware apps) and what we do (practice-aware apps). Thus, rather than providing standardised and predefined recommendations or information to all users, spatial media start from the decipherment and analysis of each individual user's preferences, activities and context, so as to subsequently provide them with personalised web content that fits best the deciphered fields of reality. In other words, with the banalisation and democratisation of new spatial media, we are moving from a universalist model of services to a model in which the basic spaces and services of everyday life increasingly become commodities that can be differentiated and adapted to the profile of each user (Graham, 2005: 565–6).

Furthermore, in terms of the power and regulatory dynamics implied by new spatial media, three other key points need highlighting. First, the surveillance-enabled personalisation of web content by new spatial media implies a regulatory dynamics – i.e. a type of governmentality in a Foucauldian sense – that does not start from a predefined normative model, but which derives recommendations through techniques of data gathering, processing and analysing, thus aiming to identify the patterns or regularities that characterise both the user's individual preferences and habits, and his or her wider context. For example, Foursquare provides users with information on specific restaurants that is not only based on the app's understanding of what places we visit and what interests we have, but also draws on a series of contextual parameters (where and when we start a restaurant search, etc.). The point is to make consumer demands and offers function better in relation to each other, thus optimising the relation between context, location and individual user needs. Reality is approached as a relationally composed whole, whose components are deciphered in their intertwined articulation. What matters is the optimised adjustment of the considered components of reality, depending on and in relation to each other.

Second, the reality-derived mode of regulation inherent in new spatial media also implies that the conveyed regulatory *telos* does not postulate a perfect and 'final' reality ever to be fully achieved, but a constant process of optimisation derived from and taking place within a given reality, whose aims and conditions are constantly re-adapted and redefined, depending not only on the perpetually changing parameters of reality itself (we have new interests, new friends, new places to visit, etc.), but also on the shifting context and conditions of regulation (new offers are available, new transport services in place, etc.). However, while new spatial media indeed aim to adapt to an ever-changing context, the images and information they convey are sometimes fixed and outdated, leading to a blurring of time references, which causes users' spatial experience to be 'continuously augmented by the "here" but not always by the "now"' (Graham et al., 2013: 477).

Third, if governing through new spatial media starts from the decoding of reality in its intertwined components, this also means that these components are not valued as either good or bad in themselves, but taken to be natural processes (in the broad sense) that are granted freedom to evolve according to their internal logics and dynamics, within the acceptable limits of the system. This implies a model of regulation and normalisation that 'work[s] within reality, by getting the components of reality to work in relation to each other, thanks to and through a series of analyses and specific arrangements … The norm is an interplay of differential normalities' (Foucault, 2007: 47, 63).

INTERCONNECTION

Importantly, spatial media work not only through the decipherment of their users' daily-life activities, but also through the interconnection of these records among each other and with other users' digital history. Many spatial media indeed offer ways to combine and store data from diverse sources. For example, as Scipioni and Langheinrich (2010: 5) have put it regarding the application Loopt, '[a] location-based recommender system has thus to match a user's individual movement history with traces from other users, find overlaps, and identify from these overlaps new places (i.e., stores, events) that the user should explore.' An identical conclusion can be drawn with regard to Foursquare's former recommendation engine Explore, which analysed users' check-in history to infer their tastes and interests. Partly based on the technique of 'collaborative filtering', Explore was recommending new places to visit on the basis of the places frequented by similar users.

In technical terms, spatial media thus exemplify the increased possibilities that now exist for interconnecting data sources situated on multiple geographical scales, and for processing and analysing in increasingly automated ways the data hence generated (Hollands, 2008). What we see emerging is a form of geographically, socially and institutionally distributed agency with regard not only to who generates data, but also to who can access the data fused and interconnected within the complex 'surveillant assemblages' (Haggerty and Ericson, 2000) underpinning everyday life. It follows that new spatial media imply a mode of regulation that aims at the perpetually more intensive and extensive study of reality, to decipher its internal regularities. We find a combined reflex towards ever more increased data gathering and ever wider circuits of data flow.

In more social terms, as techniques of 'collaborative filtering' and interconnection, new spatial media produce inherently volatile and dynamic 'filter bubbles', to use Pariser's term, standing here for the social aggregations created by the deployed computer algorithms, which contain users with similar tastes, interests and practices. Yet by rendering visible only that which corresponds to our shared tastes and interests, this 'bubbling effect' also accentuates homophilous forms of togetherness, which in turn polarises and fragments social space and indeed reinforces social exclusivity and separation (Widmer, 2015a).

ANTICIPATION

The mode of governing through personalisation and interconnection conveyed by new spatial media implies too a specific temporal logic of regulation, in which the relationship between past, present and future manifests itself in a particular way: governing relies on predefined codes, derived from the analysis of the past and applied to the present, to anticipate the future (Klauser and Albrechtslund, 2014). Thus personalised geolocational services can be considered as predictive technologies, whose functioning relies on the assumption that knowledge about the future is already present in the collected data (Amoore and De Goede, 2008). As stated by Thrift and French, 'software is deferred. It expresses the co-presence of different times, the time of its production and its subsequent dictation of future moments' (2002: 311). Algorithmic governmentality, as we see it in the case of new spatial media, is also, fundamentally, anticipatory governmentality (Amoore, 2007; Budd and Adey, 2009).

It follows that governing through new spatial media is also inherently performative in its relationship to reality. Computer algorithms constitute not only a tool of analysis but also a grammar of action (Kitchin and Dodge, 2011). As a model and technique of analysis, they simplify reality into a legible order (Budd and Adey, 2009); as a means of automated response, they perform the future through this order. In different ways and at different levels of complexity, new spatial media thus imply a relationship with reality that is at once calculated and calculating. Governing through new spatial media is both produced by and in turn produces specific classifications and orderings of reality.

One of the important questions that arises here relates to the adequacy of software to approach and govern the internal complexities and dynamics of reality. As Budd and Adey (2009: 1370) argue, 'whilst the relationship between software and the simulations they enable is often less than clear, the practice of using models and simulations is often constrained by the computing tools and languages in which they were written, limiting their accuracy and potential application.' Future research should provide more detailed empirical evidence with regard to how exactly contemporary spatial media aim to address this issue, and the wider implications this has for everyday social life.

CONCLUSION

As shown, in their reality-derived, pluralist and relative approach to reality, new spatial media aim at the surveillance-enabled provision of information and services that help manage activities, flows, etc. in highly adaptable and differentiated ways. In normative terms, the question at stake is how to know, regulate and act upon the managed reality within a 'multivalent and transformable framework' (Foucault, 2007: 20). In sum, new spatial media imply a regulatory dynamics that is fundamentally flexible and fluid in its management of reality.

This brings to the fore a fundamental conceptual and analytical problem that requires more attention in future research: how to further explore and conceptualise the fluidity and flexibility of contemporary governing through personalisation, interconnection and anticipation enabled by new spatial media. In recent years, some scholars have started to address this issue, examining the changing modalities and functioning of contemporary surveillance, from rigid and permanent monitoring and enclosure to more flexible and adaptable forms of regulation and control. Conceptually speaking, this work owes a great deal to Deleuze's (1992) essay on the 'society of control' (Boyne, 2000; Lianos, 2003; Murakami Wood, 2010), to Foucault's work centred on the concept of security (Amoore, 2006, 2011; Klauser, 2013; Klauser et al., 2014), and to Bauman's 'liquid modernity' (Bauman, 2000). As David Lyon put it in a recent conversation with Zygmunt Bauman, 'it is crucial that we grasp the new ways that surveillance is seeping into the bloodstream of contemporary life and that the ways it does so correspond to the currents of liquid modernity' (Lyon and Bauman, 2013: 152).

It would be possible and useful, we believe, to make Lyon's programmatic comment the starting point for a more sustained and systematic inquiry into the nature and functioning of contemporary software-based forms and techniques of surveillance. The three analytical axes distinguished in the present chapter – personalisation, interconnection and anticipation – could offer an initial organising framework for such inquiry.

ACKNOWLEDGEMENTS

This chapter draws upon two previously published articles: Widmer, S. (2015a) 'Experiencing a personalised augmented reality: users of Foursquare in urban space', in L. Amoore and V. Piotukh (eds), *Algorithmic Life: Calculative Devices in the Age of Big Data*, London: Routledge; and Klauser, F. and Albrechtslund, A. (2014) 'From self-tracking to smart urban infrastructures: towards an interdisciplinary research agenda on Big Data', *Surveillance and Society*, 12 (2): 273–86.

REFERENCES

Albrechtslund, A. (2012) 'Socializing the city: location sharing and online social networking', in C. Fuchs, K. Boersma, A. Albrechtslund and M. Sandoval (eds), *Internet and Surveillance: The Challenges of Web 2.0 and Social Media*. New York: Routledge. pp. 187–97.

Amoore, L. (2006) 'Biometric borders: governing mobilities in the war on terror', *Political Geography*, 25 (3): 336–51.

Amoore, L. (2007) 'Vigilant visualities: the watchful politics of the War on Terror', *Security Dialogue*, 38 (2): 215–32.

Amoore, L. (2011) 'Data derivatives: on the emergence of a security risk calculus for our times', *Theory, Culture & Society*, 28 (6): 24–43.

Amoore, L. and De Goede, M. (2008) 'Transactions after 9/11: the banal face of the preemptive strike', *Transactions of the Institute of British Geographers*, 33 (2): 173–85.

Bauman, Z. (2000) *Liquid Modernity*. Cambridge: Polity Press.

Boyne, R. (2000) 'Post-panopticism', *Economy and Society*, 29 (2): 285–307.

Budd, L. and Adey, P. (2009) 'The software-simulated airworld: anticipatory code and affective aeromobilities', *Environment and Planning A*, 41 (6): 1366–85.

Cost Action IS0807 (2008) 'Memorandum of understanding Cost Action IS0807: living in surveillance societies'. Available at: http://w3.cost.eu/fileadmin/domain_files/ISCH/Action_IS0807/mou/IS0807-e.pdf (accessed 11 August 2016).

Deleuze, G. (1992) 'Postscript on the societies of control', *October*, 53: 3–7.

de Souza e Silva, A. and Frith, J. (2012) *Mobile Interfaces in Public Spaces*. London: Routledge.

Farman, J. (2011) *Mobile Interface Theory, Embodied Space and Locative Media*. London: Routledge.

Foucault, M. (2007) *Security, Territory, Population*. London: Palgrave Macmillan.

Frith, J. (2015) *Smartphones as Locative Media*. Cambridge: Polity Press.

Graham, M., Zook, M. and Boulton, A. (2013) 'Augmented reality in urban places: contested content and the duplicity of code', *Transactions of the Institute of British Geographers*, 38 (3): 464–79.

Graham, S. (1998) 'Spaces of surveillant-simulation: new technologies, digital representations, and material geographies', *Environment and Planning D: Society and Space*, 16 (4): 483–504.

Graham, S. (2005) 'Software-sorted geographies', *Progress in Human Geography*, 29 (5): 562–80.

Green, N. and Smith, S. (2004) 'A spy in your pocket? The regulation of mobile data in the UK', *Surveillance and Society*, 1 (4): 573–87.

Haggerty, K. and Ericson, R. (2000) 'The surveillant assemblage', *British Journal of Sociology*, 51 (4): 605–21.

Hollands, R.G. (2008) 'Will the real smart city please stand up? Intelligent, progressive or entrepreneurial?', *City*, 12 (3): 303–20.

Kitchin, R. and Dodge, M. (2011) *Code/Space: Software and Everyday Life*. Cambridge, MA: MIT Press.

Klauser, F. (2013) 'Through Foucault to a political geography of mediation in the information age', *Geographica Helvetica*, 68 (2): 95–104.

Klauser, F. and Albrechtslund, A. (2014) 'From self-tracking to smart urban infrastructures: towards an interdisciplinary research agenda on Big Data', *Surveillance and Society*, 12 (2): 273–86.

Klauser, F., Paasche, T. and Söderström, O. (2014) 'Michel Foucault and the smart city: power dynamics inherent in contemporary governing through code', *Environment and Planning D: Society and Space*, 32 (5): 869–85.

Lessig, L. (2006) *Code: Version 2.0*. New York: Basic Books.

Lianos, M. (2003) 'Social control after Foucault', *Surveillance and Society*, 1 (3): 412–30.

Lyon, D. (2007) *Surveillance Studies: An Overview*. Cambridge: Polity Press.

Lyon, D. and Bauman, Z. (2013) *Liquid Surveillance*. Cambridge: Polity Press.

Mann, S., Nolan, J. and Wellman, B. (2003) 'Sousveillance: inventing and using wearable computing devices for data collection in surveillance environments', *Surveillance and Society*, 1 (3): 331–55.

Murakami Wood, D. (2010) 'Beyond the panopticon: Foucault and surveillance studies', in J. Crampton and S. Elden (eds), *Space, Knowledge and Power: Foucault and Geography*. Aldershot: Ashgate. pp. 245–64.

Pariser, E. (2011) *The Filter Bubble: What the Internet is Hiding from You*. New York: Penguin Press.

Scipioni, M. and Langheinrich, M. (2010) 'I'm here! Privacy challenges in mobile location sharing', paper presented at the *Second International Workshop on Security and Privacy in Spontaneous Interaction and Mobile Phone Use* (IWSSI/SPMU 2010), Helsinki, Finland, May 2010.

Thrift, N. and French, S. (2002) 'The automatic production of space', *Transactions of the Institute of British Geographers*, 27 (3): 309–25.

Widmer, S. (2015a) 'Experiencing a personalised augmented reality: users of Foursquare in urban space', in L. Amoore and V. Piotukh (eds), *Algorithmic Life: Calculative Devices in the Age of Big Data*. London: Routledge. pp. 57–71.

Widmer, S. (2015b) 'Navigations sur mesure ? Usages d'applications smartphone en ville de New York', *Géoregards*, 7: 55–71.

21

SPATIAL PROFILING, SORTING AND PREDICTION

DAVID MURAKAMI WOOD

INTRODUCTION

This chapter traces the rise of spatial sorting techniques from 19th-century credit reporting and mapping of social problems, through to the constant tracking and sorting conducted through networked searchable databases and geographic information systems (GIS), often termed the 'geospatial web' or simply 'geoweb' (Scharl and Tochtermann, 2007). The term 'spatial sorting' builds on David Lyon's concept of 'social sorting', a process through which data are acquired from and about individuals and used to build up profiles, which produce differential treatment in the real world (Lyon, 2001); but it emphasises that such processes are almost always spatial in nature to some degree (they differentiate places and individuals across place). Underpinned by big data, the concentration of data through cloud computing (Mosco, 2014) and expanding sources of new data, such as the 'internet of things', the geoweb is driving a diverse set of processes ranging from 'smart city' planning, to sales and marketing, home loans, credit card availability, healthcare, and now even social life and individual identity. Increasingly code, space and human subjectivity are co-constructed (Kitchin and Dodge, 2011; Murakami Wood and Ball, 2013). The approach of this chapter is to take seriously Bruce Schneier's (2012) suggestion that we understand 'big data' in the same way that we talk of 'big oil' or 'big pharma'; in other words, that attention must be paid to the political economy of 'surveillance capitalism' (Zuboff, 2015).

CREDIT BUREAUX, MARKET RESEARCH AND SEGREGATION

Commercial credit-reporting bureaux began operations in northern US cities in the mid-19th century. The Mercantile Company, which became Dun & Bradstreet,[1] was formed in 1841, and by the turn of the century companies were operating offices across the USA and in Canada (Madison, 1974). Their records allowed businesses to search for individuals on the basis of their financial behaviour; began to underpin the decision-making of the banking and insurance industry; and enabled private and state detective agencies – from Pinkerton's to the Federal Bureau of Investigation (FBI) – to sort suspects, in the latter case, because these private bureaux held vastly more data on individual Americans than state agencies (Lauer, 2010).

The application of credit-reporting techniques expanded rapidly. In the early 20th century, a new commercial discipline of market research emerged, revealing new ways of understanding desire and motivations of consumers, and how to exploit them. One of the earliest was the AC Nielsen company, which began by conducting household surveys of grocery purchases in 1923, but with the penetration of radio and then television into American households, expanded into analysis of listening and viewing habits.

From its inception such research was interested in place. Forms of classificatory social mapping emerged alongside a host of other such techniques developed by gentlemen amateur scientists, motivated by a muddle of philanthropic, welfarist, utilitarian and social Darwinist philosophies in 19th-century Paris and London. For example, Charles Booth's colour-coded streets in his 17-volume atlas of poverty (Davies, 1978) – e.g. the colour black referred to 'vicious, semi-criminal' populations – were not far removed from the early market research companies' socio-spatial classification systems. Arvidsson (2006) shows how the 'ABCD' method of simultaneously classifying groups of consumers and space combined a static idea of wealth-based class divisions, xenophobic conceptions of citizenship and belonging, and the commercial imperatives and the abilities of the marketers. According to Paul Cherrington of the J. Walter Thompson corporation, for example, Category D, was 'Homes of unskilled laborers or in foreign districts where it is difficult for American ways to penetrate' (Cherrington, 1924, quoted in Arvidsson, 2006: 49).

Racist classificatory schema could be found in the form of racially segregated state schooling and the 'redlining' of black neighbourhoods by real estate, banks and other financial services. While citizen and government action created policy changes, from the bussing of pupils to ensure mixed schools to new laws that prohibited redlining, spatial segregation and racial social sorting has persisted. For example, redlining continues to be practised, as research into so-called subprime lending that triggered the global financial crisis from 2008 indicates: 'as the patterns of foreclosures […] begin to mirror subprime activity, these vulnerabilities clearly produce racially disparate social and economic outcomes for residents of cities experiencing stress and change' (Hernandez, 2012) and 'residential segregation constitutes an important contributing cause of the current foreclosure crisis' (Rugh and Massey, 2010: 644). Similarly, the predictive policing of individuals and neighbourhoods is often racialised (Harcourt, 2006).

CLUSTER ANALYSIS AND GEODEMOGRAPHICS

Combining data on individuals and communities with mapping produced a new form of picturing populations in place. Geodemographics was pioneered by General Analytics Company, which developed multivariate cluster analysis to replace the

ABCD types of classification systems, converting official US government census data into more commercially usable forms based around the well-understood postal ZIP codes (Goss, 1995; Kaplan, 2006). Subsequently, as the Claritas corporation from 1971, the company adapted its system to the digitally encoded data that the US census began to deploy from the late 1960s (Monmonier, 2002).

Cluster analysis models were further modernised with the advent of GIS, as has the US Census Bureau from 1990 with the Topologically Integrated Geographic Encoding and Referencing (TIGER) systems.[2] Multiple specialist spatial-data companies have emerged which combine census-based geodemographic data with other data on individuals from other sources: from market surveys, credit and loyalty card records, and court records (Turrow, 2006). Some, like CACI's ACORN UK-based system[3] (see Burrows and Gane, 2006) are national specialists; others have much wider coverage.

Cluster analysis divides people in places into more sophisticated 'market segments' than ABCD classifications. Claritas's system, PRIZM, now part of the Nielsen group,[4] currently has 66 segments, produced from data classed from 'lifestage' to 'urbanicity'. A simple version available online is MyBestSegments,[5] with more sophisticated and detailed versions for paying customers, including maps down to the ZIP-code level. PRIZM is clearly hierarchical in its formulations, however light-hearted some of the designations. The segments range from 01 'Upper Crust':

> the wealthiest lifestyle in America – a haven for empty-nesting couples over the age of 55. No segment has a higher concentration of residents earning over $100,000 a year and possessing a postgraduate degree. And none has a more opulent standard of living.

Further down the scale are categories like 32 'New Homesteaders':

> Young, upper-middle-class families seeking to escape suburban sprawl find refuge in [...] a collection of small rustic townships filled with new ranches and Cape Cods. With decent-paying jobs in white and blue-collar industries, these dual-income couples have fashioned comfortable, child-centered lifestyles; their driveways are filled with campers and powerboats, their family rooms with PlayStations.

Finally, one the lowest categories, 65 'Big City Blues':

> the highest concentration of Hispanic-Americans in the nation. But it's also the multi-ethnic address for low-income Asian and African-American households occupying older inner-city apartments. Concentrated in a handful of major metros, these middle-aged singles and single-parent families face enormous challenges: low incomes, uncertain jobs, and modest educations.

As a system aimed at marketing, the classifications still stress likely purchases and consuming habits in every segment: Big City Blues folks might watch Univision and read *Star* magazine, rather than the Upper Crust's favourite Golf Channel and *The Atlantic*, but they are still assumed to be consumers.

FROM DATA TO RELATIONSHIPS

Market research was transformed again from the 1980s onward, with the rise of customer relationship management (CRM) – which advocated a more personal and trusting relationship between 'brands' and consumers created via multiple data (Murakami Wood and Ball, 2013), and algorithmic processes of dataveillance (Clarke, 1988), data mining or Knowledge Discovery in Databases (KDD). Companies like Nielsen are still essentially curators of such data, using proprietary algorithms to generate particular data sorted from their own databases.

Since the mid-1990s and accelerating in the 2000s, the proprietary model has been challenged by the rise of the internet (and particularly the World Wide Web) and another kind of proprietary algorithm: search. Although not the first, Google remains the most popular and effective internet search engine.[6] Search can be considered a form of empowerment; however, it is also part of the rise of 'personal information economies' (Elmer, 2004) and the increasing responsibilisation of citizens-turned-consumers, who have to find their own individualised answers to what were often previously considered to be social problems. As Elke Krahmann (2011) argues, 'risk discourses and practices of private businesses offer an alternative vision of the future in which industrialised societies manage their risk through individual consumer choices.' Disempowered subjects of surveillance societies themselves assess the risks posed by their fellow citizens, driven by the media and private security sectors' presentations of the worst: 'industry's ability to profit from risk management provides it with a vested interest in the creation, expansion and continuation of the demand for its services. Fear is one of its strongest marketing tools' (Krahmann, 2011: 358).

To exploit these fears, cluster analysis and GIS-based software tools have been adapted to focus on consumers' fear as much as desire, in particular, crime. Crime statistics have a long and controversial history in themselves (Haggerty, 2001); however, beginning with New York in 1996, US police departments began to adopt a new management process: CompStat (COMputer COMparison STATistics), based on mapping where crimes occurred in order to rapidly target future resources (Manning, 2008).

One of the subsequent commercial online developments of such data is Location Inc.'s Neighborhood Scout,[7] a spatial search tool that allows users to sort places based on three broad criteria – crime, schools and real estate – and provides rankings of cities. The three most popular lists at the time of writing all underlie the importance of fear: 'Most Dangerous Cities in America', 'U.S. Cities With Highest Murder Rates' and 'Safest Cities in America'. Location Inc. also provides more detailed crime-risk data by address, through its SecurityGauge service,[8] which claims to be the most spatially accurate crime data available: '48,000 × the spatial accuracy of the closest competitor', and a trustworthy prognosticator with 'Proven predictive accuracy that consistently exceeds 90%'.

However, the underlying rationale is not about people or communities at all; rather it is about risk reduction and profit maximisation for capital: 'the best location-based technologies possible to empower businesses to make informed decisions that translate to increased revenue and reduced costs'.[9] The emphasis on business empowerment and revenue indicates that individuals who use Location Inc.'s neighbourhood services are in fact less important as consumers and more important as unwitting producers of information for the higher-priced spatial search products for businesses. This process is not obvious to them as the data acquired about consumers are largely generated separately from the use of spatial search tools. But increasingly, acts of consumption are simultaneously acts of production, or 'prosumption' (Toffler, 1980). Nowhere is this clearer than on social media. The digital 'immaterial labour' (Lazzarato, 2006) involved in managing social and informational relationships creates 'overflows' (Callon, 1998), new kinds of links and connections that go beyond the simple boundaries of individual interactions, which produce a new form of surplus value for firms to exploit.

As van Dijck (2014: 198) has argued:

> [w]ith the advent of Web 2.0 and its proliferating social network sites, many aspects of social life were coded that had never been quantified before – friendships, interests, casual conversations, information searches, expressions of tastes, emotional responses, and so on.

However, it is not the overflowing data themselves that matter. Big data is already moving sorting away from older, categorical forms of sorting. Instead contemporary sorting 'privileges relational rather than demographic qualities' (Bolin and Andersson Schwarz, 2015: 1). Of particular interest are so-called 'non-obvious' relationships and what Louise Amoore (2011) has called 'data derivatives', which would not have existed without vast quantities of data and the new algorithmic tools to correlate them.

For the state, such relational data are exploited through Open Source Intelligence (OSINT) activities, or by capital through what Facebook[10] calls 'social graphing'. But there is more than simply alignment here between private marketing and state surveillance techniques. For example, the Non-Obvious Relationship Awareness (NORA) data-mining software developed by ChoicePoint, a company that had perhaps the largest American database about US voters and consumers in the mid-2000s, was used to exclude voters; and several consumer data aggregators were used by Homeland Security organisations following 9/11 in order to create new state databases (O'Harrow, 2006; Murakami Wood, 2009).[11] Nonetheless, the state still lacks the range of data collected and aggregated by private-sector data brokers.

FROM PLACE TO TRAJECTORIES

Given their connection to analysis of risks around property, it is not surprising that cluster analysis targets people and places simultaneously. However, in a mobile world, place becomes a temporary location, one of many shifting data points that define a trajectory. Such mobility is tracked primarily through mobile devices (cellphones, smartphones, laptops, tablets and the small but increasing number of wearables).

One of the primary tracking systems is another 1960s US military technology legacy, the satellite-based global positioning system (GPS), a network that enables accurate location to be calculated and tracked through the triangulation of a receiver's position relative to the constellation of satellites. Outside of its military functions, GPS is primarily used for terrain survey work, navigation in everything from container ships to private cars, and in mobile telephones and other smart devices like fitness trackers. The position data generated by GPS can be recorded and stored, enabling the tracking of the device (and indirectly, its user) across space and time. Such fine-grained tracking of location and movement has raised privacy concerns across many domains, from state surveillance to commercial profiling to household monitoring (e.g. tracking the location of children or elderly dependants) (see Chapter 22).

It is not simply that users are mobile objects generating data. Mobility has also changed the way in which spatial search data are presented to and used by consumers. Search remains central to Google's operation but has been increasingly spatialised. Google's mapping services, including Google Maps, StreetView, Google Earth and a whole range of direction-finding and location awareness and tracking services, constitute the beginning of a planetary spatial search tool. It is by no means the only such proprietary web-accessible GIS+ system, and there is also an open-source contributor-based mapping system, OpenStreetMap. There are too a growing number of location-based social media, such as Foursquare, which layer real-time personal data onto web-based mapping applications, generating location data and spatialising social connections in more conventional social media, thus creating much richer profiles on individuals.

Google has courted controversy with concerns around the capture of wireless network information by StreetView photography vehicles and private activities caught by chance on their cameras (Elwood and Leszczynski, 2011). The company addressed the latter issue with algorithms that mask faces and automobile licence plates, but these are far from infallible. Google has also experimented with combining personal location and movement data onto Google Maps, notably with its Latitude app (discontinued in 2013), which mapped contacts who had signed up for the service in real time. However, the recently implemented 'Your Timeline' feature additionally allows users to see their own movements over space and time on Google Maps. At present this can only be displayed to the user, and in order to work, the user must have location tracking enabled on their mobile device, but it is quite likely that many users will have this enabled unknowingly and are unaware their movements can be mapped. The information can undoubtedly also be data mined for commercial purposes, and British and American intelligence agencies can access Google Maps on mobile devices, alongside other eavesdropping and tracking functions, through the 'Smurf Suite' of surveillance apps (Ball, 2014).

THE POLITICAL ECONOMY OF SPATIAL SORTING

Elwood and Leszczynski remark that with the advent of the geoweb, 'privacy has gone spatial' (2011: 13). However, privacy is moving from something that could be assumed with certain exceptions, to something that is increasingly exceptional. Susan Landau (2015) has argued that Fair Information Principles (FIPs) underlying the regulation of personal data are largely incapable of dealing with the combination of big data, automation and mobility. For example, notification 'simply doesn't make much sense in a situation where collection consists of lots and lots of small amounts of information', and consent is worthless when it is an entry condition for accessing a service.

But privacy is not the only human right under attack in spatial sorting. As Manuel Aalbers notes, 'neighborhood typologies and the maps these typologies are depicted in interact with the actions of public and private actors, thereby re/producing social space' (2014: 549). For speculators, designations of vulnerability and decline may mark an opportunity for the creative destruction of gentrification and revanchist dispossession (Smith, 1996). Digitally informed rent-seeking redevelopment, combined with policy priorities that favour 'mixing', take established places built by poor, ethnic-minority and working-class communities and, in rebranding them, abstract place from people and force people away from place: a new form of digital redlining.

Mark Andrejevic has updated the ideas of the 'digital divide' to talk of 'a form of data divide not simply between those who generate the data and those who collect, store, and sort it, but also between the capabilities available to those two groups' (2014: 1674). This divide is exacerbated by its being hidden from the public gaze in the black boxes of algorithm-driven systems (Pasquale, 2015).

However, the data themselves are not simply elicited. Data are increasingly generated through the manipulation of basic emotional states (desire, fear, etc.); the information economy is also an affective economy (Thrift, 2006). The combined manipulation, measurement and application of affect exploit what would otherwise be ephemeral, unrecorded and quickly forgotten sentiments (van Dijck, 2014). Even in newly gentrified spaces, 'biopolitical marketing [...] captures the value of relatively autonomous processes of social communicative and cultural production that occur in the urban environment' (Zwick and Ozalp, 2011: 248), extracting surplus value from the prosumer behaviour of denizens. This too is often unnoticed. Shoshana Zuboff points to firms such as Google and Facebook experimenting with users' feelings in real time and that 'this new phenomenon produces the possibility of modifying the behaviours of persons and things for profit and control' (Zuboff, 2015: 85). This transformation is also visible in market moves; for example, Nielsen has acquired companies such as NeuroFocus and Innerscope, specialists in neuromarketing, which attempts to harness developments in the science of the brain to marketing (Murakami Wood and Ball, 2013).

Such power demands accountability and regulation, yet Zuboff shows that Google rejects such demands, even though their products function as essential public utilities. The situation is profoundly different from earlier forms of capitalism in which capitalists, although powerful, could be brought down: 'Google', she says, 'bears no such risks' (Zuboff, 2015: 88). On the contrary, the relationship between capital and state is one in which risks are transferred from capital to the state, and in which individual risk is opportunity: as Elke Krahmann argues, 'the fact that risks can never fully be eliminated promises insatiable demand' (2011: 357).

Moreover, the state seemingly has little interest in assisting citizens in holding information corporations to account. As van Dijck (2014, original emphasis) points out, 'the institutions gathering and processing big data are not organized *apart from* the agencies that have the political mandate to regulate them'. A close relationship between technology producers, private data aggregators and public administration has been particularly central to the hegemonic American neoliberal model of capitalism. The state reduces, manages and compensates for the risks inherent in investment decisions, and in return, it seems, corporations cooperate with the state in providing user data, as all the largest information corporations do as part of the National Security Agency's (NSA) PRISM programme (Greenwald, 2014).[12]

However, the diffusion of anticipatory surveillance throughout state, corporate and personal action means that any central locus of power may be, like the individual capitalist, spectral. Palmås (2011), drawing on De Landa (1991), calls this 'panspectric' rather than panoptic surveillance, and Zuboff's (2015: 80) recent analysis of 'surveillance capitalism' agrees, asking not only that we replace Bentham's Panopticon in our analysis, but also that we replace the figure of Orwell's Big Brother with a more diffuse but pervasive, 'Big Other':

> habitats inside and outside the human body are saturated with data and produce radically distributed opportunities for observation, interpretation, communication, influence, prediction, and ultimately modification of the totality of action. Unlike the centralized power of mass society, there is no escape from Big Other. There is no place to be where the Other is not.

Dan Trottier (2014: 69), though, gives some hope in the incompleteness of any such process, and the very human structures in which activities are still embedded: 'while large scale data monitoring is possible, such practices are shaped by situated cultures and material constraints'; and it is probably too early to judge whether the results of current efforts to regulate and control, 'fledgling agencies and legal uncertainties are grounds for emerging configurations and unanticipated hazards'. It seems a poor sort of hope. However, in the recognition of the political economic forces driving ubiquitous surveillance, there is at least the possibility for action.

NOTES

1. Dun & Bradstreet remains a large business analytics company, with a market capitalisation of just short of US$4 billion (all market data in this chapter were obtained from Reuters data via Google Finance, 1 November 2015).
2. This remains the basis for the geographical sorting and representation of US census data; see: http://quickfacts.census.gov/qfd/meta/long_LND110210.htm
3. ACORN website: http://acorn.caci.co.uk/
4. Nielsen was bought and then split up and sold by Dun & Bradstreet. In 2006 it was reformed, backed by private equity from a consortium including the Carlyle Group, and expanded by acquiring Marketing Analytics and public opinion-polling organisation the Harris Group. Nielsen is not a publicly traded company, but it was rated as the world's largest market research company in the Honomichl 2014 rankings from the American Marketing Association.
5. PRIZM ('MyBestSegments') website: https://segmentationsolutions.nielsen.com/mybestsegments/Default.jsp
6. Emerging from military research-funded work at Stanford University in the early–mid-1990s, Google has both innovated and acquired over 180 smaller companies to become one of the biggest information-based businesses in the world. Alphabet, Google's new parent company from 2015, has a market capitalisation of over US$510 billion; only Apple is bigger at around US$670 billion.
7. http://www.neighborhoodscout.com/
8. http://www.securitygauge.com/
9. http://www.locationinc.com/about-us
10. Currently the largest social media provider in the world, with a market capitalisation of some US$310 billion.

11. In 2008, ChoicePoint was acquired by transnational information aggregator Reed Elsevier, which has since become RELX Group, a growing company with a market capitalisation of nearly US$20 billion.

12. The NSA's PRISM programme has no relationship to Nielsen's PRIZM software.

REFERENCES

Aalbers, M. (2014) 'Do maps make Geography? Part 1: Redlining, planned shrinkage, and the places of decline', *ACME: An International E-Journal for Critical Geographies*, 13 (4): 525–56.

Amoore, L. (2011) 'Data derivatives: on the emergence of a security risk calculus for our times', *Theory, Culture & Society*, 28 (6): 24–43.

Andrejevic, M. (2014) 'The big data divide', *International Journal of Communication*, 8: 1673–89.

Arvidsson, A. (2006) *Brands: Meaning and Value in Media Culture*. London: Routledge.

Ball, J. (2014) 'Angry Birds and "leaky" phone apps targeted by NSA and GCHQ for user data', *Guardian*, 28 January. Available at: http://www.theguardian.com/world/2014/jan/27/nsa-gchq-smartphone-app-angry-birds-personal-data (accessed 27 July 2016).

Bolin, G. and Andersson Schwarz, J. (2015) 'Heuristics of the algorithm: big data, user interpretation and institutional translation', *Big Data and Society*, 2 (2): 1–12.

Burrows, R. and Gane, N. (2006) 'Geodemographics, software and class', *Sociology*, 40 (5): 773–91.

Callon, M. (1998) *The Laws of the Markets*. Oxford: Blackwell.

Clarke, R. (1988) 'Information technology and dataveillance', *Communications of the ACM*, 31 (5): 498–512.

Davies, W.K.D. (1978) 'Charles Booth and the measurement of urban social character', *Area*, 10 (4): 290–6.

De Landa, M. (1991) *War in the Age of Intelligent Machines*. Cambridge, MA: Zone Books.

Elmer, G. (2004) *Profiling Machines: Mapping the Personal Information Economy*. Cambridge, MA: MIT Press.

Elwood, S. and Leszczynski, A. (2011) 'Privacy, reconsidered: new representations, data practices, and the geoweb', *Geoforum*, 42: 6–15.

Goss, J. (1995) '"We know who you are and we know where you live": the instrumental rationality of geodemographic systems', *Economic Geography*, 71 (2): 171–98.

Greenwald, G. (2014) *No Place to Hide: Edward Snowden, the NSA, and the US Surveillance State*. London: Macmillan.

Haggerty, K.D. (2001) *Making Crime Count*. Toronto: University of Toronto Press.

Harcourt, B.E. (2006) *Against Prediction: Profiling, Policing and Punishing in an Actuarial Age*. Chicago, IL: Chicago University Press.

Hernandez, J. (2012) 'Redlining revisited: mortgage lending patterns in Sacramento 1930–2004', in Aalbers, M. (ed.), *Subprime Cities: The Political Economy of Mortgage Markets*. Chichester: Wiley. pp. 187–217.

Kaplan, C. (2006) 'Precision targets: GPS and the militarization of US consumer identity', *American Quarterly*, 58 (3): 693–714.

Kitchin, R. and Dodge, M. (2011) *Code/Space: Software and Everyday Life*. Cambridge, MA: MIT Press.

Krahmann, E. (2011) 'Beck and beyond: selling security in the world risk society', *Review of International Studies*, 37 (1): 349–72.

Landau, S. (2015) 'Control use of data to protect privacy', *Science*, 347: 6221.

Lauer, J. (2010) 'The good consumer: credit reporting and the invention of financial identity in the United States, 1840–1940', *Enterprise and Society*, 11 (4): 686–94.

Lazzarato, M. (2006) 'Immaterial labour', in P. Virno and M. Hardt (eds), *Radical Thought in Italy: A Potential Politics*. Minneapolis, MN: University of Minnesota Press. pp. 133–47.

Lyon, D. (2001) *Surveillance Society: Monitoring Everyday Life*. Milton Keynes: Open University Press.

Madison, J.H. (1974) 'The evolution of commercial credit reporting agencies in nineteenth-century America', *Business History Review*, 48 (2): 164–86.

Manning, P.K. (2008) *The Technology of Policing: Crime Mapping, Information Technology, and the Rationality of Crime Control*. New York: NYU Press.

Monmonier, M. (2002) *Spying with Maps: Surveillance Technologies and the Future of Privacy*. Chicago, IL: University of Chicago Press.

Mosco, V. (2014) *To the Cloud: Big Data in a Turbulent World*. New York: Paradigm.

Murakami Wood, D. (2009) 'Spies in the information economy: academic publishers and the trade in personal information', *ACME: An International E-Journal for Critical Geographies*, 8 (3): 484–93.

Murakami Wood, D. and Ball, K.S. (2013) 'Brandscapes of control? Surveillance, marketing and the co-construction of subjectivity and space in neo-liberal capitalism', *Marketing Theory*, 13 (1): 47–67.

O'Harrow, R. (2006) *No Place to Hide*. New York: Simon & Schuster.

Palmås, K. (2011) 'Predicting what you'll do tomorrow: panspectric surveillance and the contemporary corporation', *Surveillance and Society*, 8 (3): 338–54.

Pasquale, F. (2015) *The Black Box Society: The Secret Algorithms that Control Money and Information*. Cambridge, MA: Harvard University Press.

Rugh, J.S. and Massey, D.S. (2010) 'Racial segregation and the American foreclosure crisis', *American Sociological Review*, 75 (5): 629–51.

Scharl, A. and Tochtermann, K. (2007) *The Geospatial Web: How Geobrowsers, Social Software and the Web 2.0 are Shaping the Network Society*. London: Springer.

Schneier, B. (2012) *Liars and Outliers: Enabling the Trust That Society Needs to Thrive*. New York: Wiley.

Smith, N. (1996) *The New Urban Frontier: Gentrification and the Revanchist City*. New York: Psychology Press.

Thrift, N. (2006) 'Re-inventing invention: new tendencies in capitalist commodification', *Economy and Society*, 35 (2): 279–306.

Toffler, A. (1980) *The Third Wave*. New York: Bantam.

Trottier, D. (2014) 'Big data ambivalence: visions and risks in practice', in M. Hand and S. Hillyard (eds), *Big Data? Qualitative Approaches to Digital Research*. Bingley: Emerald. pp. 51–72.

Turrow, J. (2006) *Niche Envy: Marketing Discrimination in the Digital Age*. Cambridge, MA: MIT Press.

van Dijck, J. (2014) 'Datafication, dataism and dataveillance: Big Data between scientific paradigm and ideology', *Surveillance and Society*, 12 (2): 197–208.

Zuboff, S. (2015) 'Big Other: surveillance capitalism and the prospects of an information civilization', *Journal of Information Technology*, 30: 75–89.

Zwick, D. and Ozalp, Y. (2011) 'Flipping the neighbourhood: biopolitical marketing as value creation for condos and lofts', in D. Zwick and J. Cayla (eds), *Inside Marketing: Practices, Ideologies, Devices*. Oxford: Oxford University Press. pp. 234–53.

22
GEOPRIVACY
AGNIESZKA LESZCZYNSKI

INTRODUCTION

This chapter examines the ways in which the rapid proliferation and resulting pervasiveness of spatial media are radically reconfiguring norms and expectations around locational privacy. Over the past decade, there has been an extensive commercialisation of all things 'geo' (Wilson, 2012; Leszczynski, 2014). This may be evidenced in the ubiquity and ordinariness of locationally enabled devices, mapping platforms, spatial interfaces, geosocial applications and myriad location-based services in the spaces and practices of the everyday. Many of our quotidian digital media practices are spatially oriented. They depend on the availability of geocoded information as functional inputs to the applications and services that we regularly use. We generate spatial content as intended outputs or byproducts of our interactions with spatial media, at times unbeknown to us. Locational affordances such as global positioning systems (GPS), Wi-Fi, Bluetooth and gyroscopes are now standard features of most digital devices. Our mobile devices, operating systems and applications log and transmit our personal spatial information as data, sometimes unencrypted and passed in the clear. Many of the applications and services installed on our devices ask for permission to access these locational affordances upon installation, or harness them in the background without our knowledge. Much of the digital content we produce and share is or may easily be geocoded. The rise of crowdsourcing allows us to generate geodata not only about ourselves but also about others, with or without their consent (Ricker et al., 2014). Simultaneously, as we move through the urban fabric, we and the devices we carry on our persons register our time-stamped presences as data events against myriad sensors distributed across the extensively monitored landscapes of the smart city (Kitchin, 2015). Elsewhere, our presences are captured as temporally decontextualised visual elements of commercial spatial fabrics, assembled by corporate giants such as Google (StreetView) and Microsoft (Bing StreetSide) (Elwood and Leszczynski, 2011).

Locational privacy has until quite recently been defined in terms of societal norms latent in the presumption that as we move through the material spaces of our daily lives, our locations are not being surreptitiously and systematically monitored, recorded, stored and later repurposed in ways that are compromising of our safety, security and/or confidentiality (Blumberg and Eckersley, 2009). Yet in the networked locational data and device ecologies of spatial media and the parallel realities of living under conditions of continuous geosurveillance (Kitchin, 2015), these presumptions no longer hold. With spatial media, our movements, behaviours, and actions in, through and across space are easily and seamlessly digitally generated, captured, registered, leaked, intercepted, transmitted, disclosed, dis/assembled across data streams, to be repurposed by ourselves and others. Our personal spatial data flow freely and without friction across and between interoperable and synergistic geo-enabled devices, platforms, services, applications, and analytics engines.

Existing definitions of locational privacy such as the one offered above are individualistic, emphasising a negatively defined rights-oriented approach to privacy – for example, the right not to have one's location monitored and recorded when going about quotidian activities in space. Privacy, however, is being relocated from the individual to the network, where privacy violations and harms increasingly occur beyond the site of the individual. For example, not only do privacy harms result from the capture of an individual's movements as data, but also their disclosure, sale, repurposing, and analytics by subsequent parties (Marwick and boyd, 2014). In the networked ecologies of digital practice, controlling the flow of one's personal data is difficult, impractical, and arguably unfeasible (Marwick and boyd, 2014; The White House, 2014; Leszczynski, 2015).

Privacy harms and violations do not arise solely from the disclosure of individuals' locations, or from their being placed on a map. New possibilities for privacy harm and violation are presented by the inherent relationality of spatial data as well as the spatio-temporal nature of geodata events. Spatio-temporal data – data that include spatial as well as temporal referents – allow for the tracking and reconstruction of not only position but also movement of individuals. The intrinsic relationality of big data phenomena (boyd and Crawford, 2012) means that individuals' spatial information may be easily correlated with other kinds of personally identifying information (PII). For example, it may be used to infer political, social and/or religious affiliation based on co-proximity and co-movement with others, revealing membership in particular groups (Soltani and Gellman, 2013). The relational nature of geodata furthermore presents new possibilities for privacy violation stemming from the ways in which location is functionally synonymous with, and a data proxy for, identity (de Montjoye et al., 2013). Simultaneously, individuals' locations may be algorithmically inferred to a high degree of accuracy from other digital metadata (e.g. temporal referents), removing the necessity of direct locational data disclosure on the part of individuals themselves (Priedhorsky et al., 2014).

Encompassing more than solely location, then, a broadened concept of 'geoprivacy' must account for the emergent complex of potential privacy harms and violations that may arise from a number of nascent realities of living in a (spatial) big data present:

1. from the spatial-media-enabled pervasive capture and repurposing of individuals' personal spatial-relational and spatio-temporal data;

2. from the ways in which individuals cast digital footprints as they move across the numerous sensor networks of smart cities;

3. from the circulation and analytics of these data, which position individuals as spatially precarious in various and unprecedented ways; and

4. from the inability of individuals to control highly personal flows of spatial information about themselves in networked device and data ecologies.

Networked data and device ecologies are not unique to spatial media, but characterise much of mundane everyday digital practices and digitally mediated interactions with others (persons, content and hardware/software objects). As compared to other kinds of PII, however, geolocation is uniquely sensitive in terms of the kinds of information it can reveal about individuals, and the ways in which those disclosures are made. Moreover, the contexts of the capture, circulation and repurposing of individuals' spatial data are: (1) distributed (occur across multiple devices, applications and services), (2) platform-independent (data flow easily across platforms, services and devices) and (3) indiscriminate (involve potentially all individuals). This makes the fashioning of practices and tactics for evading geosurveillance, or for seeking obscurity and anonymity within data flows, far more difficult when compared to other forms of digital tactics, particularly those that coalesce around dominant social platforms such as Facebook or Twitter, such as disabling and reactivating one's Facebook profile in between uses of the platform, for example (boyd, 2014).

The particular sensitivity of individuals' personal spatial information and the complexity of exerting control over flows of these data make geoprivacy uniquely deserving of attention. In the sections that follow, I account for some of the multiple ways in which individuals become abstracted as data events into spatial big data flows by virtue of simply going about their daily lives. I subsequently identify both why spatial data are uniquely sensitive (in terms of the kinds of things that they can be used to reveal about individuals), and the ways in which such disclosures are made. This sensitivity, as well as the ways in which personal spatial data capture and repurposing positions individuals as spatially precarious in different ways, make geoprivacy a prominent concern within broader societal debates about privacy and digitality. Yet, the continuous, extensive, cross-platform and non-selective nature of geodata capture, collection, mining, interception, and analytics presents particular challenges for controlling flows of personal spatial information in networked data and device ecologies.

NETWORKED ECOLOGIES OF SPATIAL BIG DATA

A total of 62% of UK adults own a smartphone (Ofcom, 2014); in the US, 64% do (Pew Research Center, 2015). These devices ooze location. They ping off cell towers when they come within range, registering their time-stamped presence.

Their built-in GPS functionality allows user location to be pinpointed with a high degree of accuracy and logged as such within mobile operating systems, harnessed by applications and services, and attached as metadata to user-generated content. Whenever Wi-Fi is enabled, devices continuously look for networks to join, and in so doing, can easily be registered as time-stamped data events when recognised by the router. Oftentimes, these are networks that enable legitimate internet access, but this also includes network emulators that force communication with smart-phones, recording which unique devices appear on the network while potentially also syphoning additional data from these devices, including locational data logs. In 2013, for example, the Seattle Police Department purchased a Wi-Fi mesh to privilege emergency response and relief communications in the event of an urban disaster, and subsequently applied to the city attorney's office to use the network as an everyday surveillance mesh (Hamm, 2013). The mesh does not grant individual access onto any useable network; however, whenever devices come within range of any of the 160 network access points (NAPs) distributed across the city, the access points show up as available networks for users to join. Upon recognising a device as within range, the mesh is capable of storing unique device IDs, the device type, applications installed on the device, as well as the locational history for the last 1000 times that a unique device has emitted a signal to the network (Hamm, 2013).

It is not only state entities that participate in these emergent practices of geo-surveillance, but corporate actors as well. For instance, likewise in 2013, a private company installed a series of Wi-Fi-enabled bins with digital advertising displays along a busy shopping street in London to track foot traffic patterns and target advertising to individual pedestrians, identified on the basis of unique device IDs and the spatio-temporal patterns of their passing by any of these bins (Datoo, 2013). At even closer range, smartphones may be spatially tracked, for instance in retail environments, to within mere inches away from iBeacon receivers that exploit the Bluetooth functionality of mobile devices.

In some instances, such exploitation of the locational affordances of devices is functionally necessary to the use of an application or service; for example, the social recommendation service Foursquare harnesses the real-time locational signal of users' devices to determine nearby venues (for eating, shopping, drinking and taking in cul-tural events) to be pushed out as notifications to users' mobiles. Elsewhere, however, applications veil locational data-mining operations behind primary interfaces – such as mobile games – that are not functionally reliant upon location, but collect this information to sell on to subsequent vendors. An analysis of the 50 most popular applications installed on Android and iOS devices, respectively, conducted by the *Wall Street Journal* in 2011 revealed that of the 50 iPhone apps analysed, 50% transmitted location data to a third party other than the app developer. Of these, a full 19 were neither location-based services nor provided native locational functionality. Similarly 21 of the top 50 Android apps were identified as disseminating user location data beyond the app developers; for 13 of these, accessing the locational capabilities of the user's device is not an identifiably functional requisite for the app (*The Wall Street Journal*, 2011). Fifty per cent of iOS apps passing locational data on to a third party did not request user permission to do so; a third of Android apps similarly did not request consent (*The Wall Street Journal*, 2011). These applications and services have been revealed to be 'leaky'; that is, they pass individuals' spatial data, at times unencrypted, to mobile advertisers, allowing for easy interception by other parties, including the securities services (Glanz et al., 2014).

The examples presented above capture only some of the ways in which individuals' geolocation data are generated, captured, shared, leaked, and intercepted. What is significant about these new forms of personal spatial–data disclosure is that they position individuals as spatially precarious in new and unprecedented ways. By spatial precariousness, I mean to designate the ways in which individuals become exposed or predisposed to experiencing material and personal harms – such as physical violence – in space (Leszczynski and Elwood, 2015). The generation, capture, sharing, dissemination and repurposing of personal spatial big data flows create potentials for either making individuals spatially vulnerable or for heightening existing levels of spatial vulnerability, by virtue of geoprivacy harms and violations that arise from the unique sensitivity of personal spatial–information disclosure and also the difficulties of controlling flows of personal geolocation information in networked data and device ecologies.

WHAT MAKES LOCATION DATA SO SENSITIVE?

There are four identifiable ways in which personal spatial data are uniquely sensitive in terms of the kinds of things that they can be used to reveal about individuals, and the ways in which those revelations are made:

1. Spatio-temporal location is seen to constitute definitive proof or evidence of individuals' involvement in specific behaviours, activities and events in space, or is seen as proof of the potential of their involvement.

2. The extensive, exhaustive and continuous nature of geosurveillance (Kitchin, 2015) means that there is no feasible way of achieving or maintaining complete spatial anonymity within data flows.

3. Spatial–relational data are inherently meaningful beyond being merely locational, revealing other intimate aspects of our personal lives.

4. Unlike other forms of PII, spatial data carry with them information that can be used to translate threats to our personal safety and security into actual harms to our person.

In 2009, an image captured from Google StreetView in Switzerland made the news. It showed a married Swiss politician walking down the street with a woman who was not his wife, with the suggestion being that he was involved in an extra-marital affair (Klapper, 2009). The woman was actually a member of his staff; the image showed no inappropriate behaviour or interaction between these two individuals. This example is, however, telling in that the visual orientation of this spatial medium – Google StreetView – seemingly constitutes definitive proof of a person's complicity in socially condemnable behaviour. The purported legitimacy of the evidence is latent, in this instance, in the simultaneously photorealistic and spatial nature of the disclosure (see Elwood and Leszczynski, 2011). In the big data present, spatial media are increasingly being enrolled not only as seemingly definitive proof of our actions, behaviours and movements in place, but also of the *potential* of our complicity in social and political events and activities in various ways.

On the basis of the Snowden revelations, we know that spatial data are figuring as increasingly important in the emergent intelligence activities of the Western security agencies, as new surveillance regimes are crystallising around big data and analytics. A particular suite of tools developed by the National Security Agency (NSA), code-named co-traveler, designates a number of analytics techniques for the bulk-process-ing of cellphone metadata, of which the NSA collect 5 billion a day worldwide, to identify individuals of potential interest based on locational behaviours they exhibit, regardless of whether they were previously a suspect or not (Soltani and Gellman, 2013). This analytics suite has been designated 'co-traveler' because it allows for the identification of co-presence and co-movement. If an individual's mobile device pings off the same cell towers in the same sequence as that of an existing target of surveillance, then they may be considered as being a 'co-traveller', similarly involved in terrorist-related activity, for example.

The ability for media reporters or the securities agencies to pinpoint our move-ments in time and space carries weight and legitimacy as proof of our complicity in certain activities or culpability in certain kinds of events, even where these data are inaccurate (as in the case of the Swiss politician) or coincidental (how many cell-tower 'pings' in sequence are indicative of co-travel?). The ability of the security services to capture and analyse personal spatial data on such a broad, unprecedented scale is significant because it means that they are 'able to render most efforts at [securing communications] effectively futile' (Gellman and Soltani, 2013, no page). This dragnet nature of geosurveillance means that measures taken to evade surveil-lance, seek obscurity within data flows or conceal identity are effectively meaning-less because an individual cannot possibly achieve anonymity/obscurity/evasion in all possible contexts. It has recently been demonstrated that even with all locational affordances disabled on a device, a smartphone's power-usage record may be used as a proxy for location, as power usage is greater the further a device is away from a cell tower (Kleinman, 2015). Even were an individual to forego ownership of a digital device and suspend all means of digital communication – an impractical-ity and unreasonable expectation in the pervasively digitally mediated spaces and practices of everyday life – their quotidian spatial movements would continue to be abstracted into spatial big data flows through capture by sensor networks such as CCTV with automated facial recognition systems.

Geolocation is of great interest to state actors and corporate entities because spatial data are inherently meaningful. The kinds of places we visit are innately revealing of other facets of our identity that would in other contexts be considered private and enshrined ethically and legally as information that should never be disclosed. For example, a mobile app mining device location data could easily track a user's visits to specialist medical offices and discern the nature of the speciality (Dwoskin, 2014), even pinpointing device location to a specific office within a building via Indoor Positioning System (IPS) technology (DesMarais, 2012). We collectively understand our genetic information to be private and something that should never be disclosed to an insurance company. But what happens if a mobile app discerns an individual making repeated visits to a medical professional specialising in a degenerative disease, and sells these data on to an insurance company? Spatial data are as sensitive as our confidential medical information because of the things that they can reveal about us (such as our health status), and they do so in ways that circumvent social norms, expectations, and legal privacy protections. Not only may intimate aspects of our identity be discerned from our spatial data, but actually our personal spatial data *are* our identity. Indeed, a group of researchers determined that unique individuals could be identified from the spatial metadata of only four cellphone calls (de Montjoye et al., 2013).

Personal geolocation raises unique privacy concerns because it carries with it material information that may be used to actualise threats to our personal safety and security. Longitudinal spatial data tracking and dissemination allow for the reconstruction of individuals' detailed spatial histories. In 2011, it was revealed that Apple mobile devices had been logging up to a year's worth of highly precise, time-stamped device location data, storing this information in an unencrypted file within the operating system (iOS) itself, even where users had intentionally disabled location services on their devices (Arthur, 2011). This creates opportunities for other individuals to discern where a person is most likely to be on a weekday afternoon between the hours of one and two, for example. While this scenario may be a highly unlikely one, requiring both intent and technical knowledge, concerns over digital, spatial data-enabled stalking are not unwarranted. In the 12 months leading up to April 2014, UK police received 10,731 reports of individuals having their devices compromised via malicious spyware and malware installed for the express purposes of digitally stalking them, including by tracking their locations (Doward, 2015).

Elsewhere, the capacities for the repurposing of personal spatial data through public application programming interfaces (APIs) may likewise be used to predispose select social groups to physical harm. In 2012, the popular geosocial application Foursquare revoked access to its API by the application 'Girls Around Me' (GAM), which scraped public check-ins for those generated by users with female-sounding first names and correlating them with these users' public Facebook profiles (Brownlee, 2012). This created possibilities for the actualisation of predatory behaviour, providing users with information about places where particular women may be present, and enabling use of highly personal details (likes and dislikes in music, movies and books, as well as their relationship status) discerned from their social profiles to open a conversation and establish false commonality of interests with the intention of, hypothetically, organising and enacting a sexual assault or unsolicited sexual advances.

CHALLENGES FOR GEOPRIVACY IN NETWORKED ECOLOGIES OF SPATIAL MEDIA

The GAM example is particularly salient because it captures the immediate challenges of controlling flows of personal locational information in the networked ecologies of spatial media. That women's Foursquare check-ins were being mined, correlated with their social media profile information, and used to effectively position them as spatially vulnerable targets of sexual overtures was undoubtedly largely unbeknown to the vast majority of female users whose locations were advertised through the third-party application because GAM never asked them for consent – it simply assumed it. This is possible given the networked nature of platforms in which users can log into one service (e.g. Foursquare) using their authentication credentials from another service (e.g. Facebook), as well as the platform-independence of geolocational content itself, which means that, once generated, it circulates freely and is easily repurposed by subsequent parties towards uses for which we never intended, and often without our knowledge or consent. As Ricker et al. (2014) argue, data and their subsequent disclosure to other parties 'is an interconnected cyclical process making it difficult to distinguish when and where each act [of data collection, disclosure, access, etc.] begins and ends'.

In the emergent networked ecology of data and devices, exerting individual control over the flow of personal information, including location data, is extremely difficult and often highly impractical (Marwick and boyd, 2014). However, the commercial control of platforms, operating systems, devices, services, and applications means that individuals do not define the contexts of information flow (Nissenbaum, 2010). Even where there are attempts to architecture in privacy by design, privacy settings tend to be overwhelmingly binaristic, permitting either full disclosure (public; broadcast to the world) or complete privacy (permissions need to be granted on an individual level) (Elwood and Leszczynski, 2011; Marwick and boyd, 2014). This is incommensurate with how we participate on digital platforms, in which we want to be able to share different information with different parties at varying levels of disclosure (Marwick and boyd, 2011, 2014). The significance of this is that our conceptualisation of geoprivacy needs to move away from highly axiomatic definitions that stress the individual and a negative definition of privacy rights ('freedom from'), to one that encompasses and accounts for the realities of continuous personal locational data flow as a feature of everyday digital practices that are characterised by extensive, real-time geosurveillance and the networked data and device ecologies of spatial media.

REFERENCES

Arthur, C. (2011) 'iPhone keeps record of everywhere you go', *Guardian*, 20 April. Available at: http://www.guardian.co.uk/technology/2011/apr/20/iphone-tracking-prompts-privacy-fears (accessed 28 July 2016).

Blumberg, A.J. and Eckersley, P. (2009) 'On locational privacy, and how to avoid losing it forever', white paper, Electronic Frontier Foundation. Available at: http://www.eff.org/wp/locational-privacy (accessed 28 July 2016).

boyd, d. (2014) *It's Complicated: The Social Lives of Networked Teens.* New Haven, CT: Yale University Press.

boyd, d. and Crawford, K. (2012) 'Critical questions for big data', *Information, Communication and Society*, 15 (5): 622–79.

Brownlee, J. (2012) 'This creepy app isn't just stalking women without their knowledge, it's a wake-up call about Facebook privacy', *Cult of Mac*, 30 March. Available at: http://www.cultofmac.com/157641/this-creepy-app-isnt-just-stalking-women-without-their-knowledge-its-a-wake-up-call-about-facebook-privacy/ (accessed 28 July 2016).

Datoo, S. (2013) 'This recycling bin is stalking you', *The Atlantic Cities*, 8 August. Available at: http://www.theatlanticcities.com/politics/2013/08/recycling-bin-following-you/6475/ (accessed 28 July 2016).

de Montjoye, Y.A., Hidalgo, C.A., Verleysen, M. and Blondel, V.D. (2013) 'Unique in the crowd: the privacy bounds of human mobility', *Scientific Reports*, 3 (1376).

DesMarais, C. (2012) 'This smartphone tracking tech will give you the creeps', *PC World*, 22 May. Available at: http://www.pcworld.com/article/255802/new_ways_to_track_you_via_your_mobile_devices_big_brother_or_good_business_.html (accessed 28 July 2016).

Doward, J. (2015) 'Spyware and malware availability sparks surge in internet stalking', *Guardian*, 28 February. Available at: http://www.theguardian.com/society/2015/feb/28/spyware-malware-surge-internet-stalking-digital-trust-gps (accessed 28 July 2016).

Dwoskin, E. (2014) 'What secrets your phone is sharing about you', *The Wall Street Journal*, 13 January. Available at: http://online.wsj.com/news/articles/SB10001424052702303453004579290632128929194 (accessed 28 July 2016).

Elwood, S. and Leszczynski, A. (2011) 'Privacy, reconsidered: new representations, data practices, and the geoweb', *Geoforum*, 42 (1): 6–15.

Gellman, B. and Soltani, A. (2013) 'NSA tracking cellphone locations worldwide, Snowden documents show', *The Washington Post*, 4 December. Available at: http://www.washingtonpost.com/world/national-security/nsa-tracking-cellphone-locations-worldwide-snowden-documents-show/2013/12/04/5492873a-5cf2-11e3-bc56-c6ca94801fac_story.html (accessed 28 July 2016).

Glanz, J., Larson, J. and Lehren, A.W. (2014) 'Spy agencies tap data streaming from phone apps', *The New York Times*, 27 January. Available at: http://www.nytimes.com/2014/01/28/world/spy-agencies-scour-phone-apps-for-personal-data.html?_r=0 (accessed 28 July 2016).

Hamm, D. (2013) 'Seattle police have a wireless network that can track your every move', Kirotv.com, 23 November. Available at: http://www.kirotv.com/news/news/seattle-police-have-wireless-network-can-track-you/nbmHW/ (accessed 28 July 2016).

Kitchin, R. (2015) 'Continuous geosurveillance in the "smart city"', *DIS Magazine*, February. Available at: http://dismagazine.com/issues/73066/rob-kitchin-spatial-big-data-and-geosurveillance/ (accessed 28 July 2016).

Klapper, B.S. (2009) 'Swiss official tells Google to erase street views', phys.org, 24 August. Available at: phys.org/pdf170330788.pdf (accessed 3 August, 2016).

Kleinman, Z. (2015) 'Battery power alone can be used to track Android phones', BBC News, 23 February. Available at: http://www.bbc.com/news/technology-31587621 (accessed 28 July 2016).

Leszczynski, A. (2014) 'On the neo in neogeography', *Annals of the Association of American Geographers*, 104 (1): 60–79.

Leszczynski, A. (2015) 'Spatial big data and anxieties of control', *Environment and Planning D: Society and Space*, 33 (6): 965–84.

Leszczynski, A. and Elwood, S. (2015) 'Feminist geographies of new spatial media', *The Canadian Geographer*, 59 (1): 12–28.

Marwick, A.E. and boyd, d. (2011) 'I tweet honestly, I tweet passionately: Twitter users, context collapse, and the imagined audience', *New Media and Society,* 13 (1): 96–113.

Marwick. A.E. and boyd, d. (2014) 'Networked privacy: how teenagers negotiate context in social media', *New Media and Society*, 16 (7): 1051–67.

Nissenbaum, H.F. (2010) *Privacy in Context: Technology, Policy, and the Integrity of Social Life*. Palo Alto, CA: Stanford Law Books.

Ofcom (2014) 'Adults' media use and attitudes report 2014'. Available at: http://stakeholders.ofcom.org.uk/market-data-research/other/research-publications/adults/adults-media-lit-14/ (accessed 28 July 2016).

Pew Research Center (2015) 'The smartphone difference'. Available at: http://www.pewinternet.org/2015/04/01/us-smartphone-use-in-2015/ (accessed 28 July 2016).

Priedhorsky, R., Culotta, A. and Del Valle, S.Y. (2014) 'Inferring the origin locations of tweets with quantitative confidence', *Proceedings of the 17th ACM conference on Computer Supported Cooperative Work and Social Computing (CSCW'14)*. New York: ACM. pp. 1523–36.

Ricker, B., Schuurman, N. and Kellser, F. (2014) 'Implications of smartphone usage on privacy and spatial cognition: academic literature and public perceptions', *GeoJournal*, 80 (5): 637–52.

Soltani, A. and Gellman, B. (2013) 'New documents show how the NSA infers relationships based on mobile location data', *The Switch*, 10 December. Available at: http://www.washingtonpost.com/blogs/the-switch/wp/2013/12/10/new-documents-show-how-the-nsa-infers-relationships-based-on-mobile-location-data/ (accessed 28 July 2016).

The Wall Street Journal (2011) 'What they know – mobile'. Available at: http://blogs.wsj.com/wtk-mobile/ (accessed 28 July 2016).

The White House (2014) 'Big data: seizing opportunities, preserving values'. Washington, DC: Executive Office of the President. Available at: https://www.whitehouse.gov/sites/default/files/docs/big_data_privacy_report_may_1_2014.pdf (accessed 11 August 2016).

Wilson, M.W. (2012) 'Location-based services, conspicuous mobility, and the location-aware future', *Geoforum*, 43 (6): 1266–75.

INDEX